The Essentials of Telecommunications Management

A Simple Guide to Understanding a Complex Industry

Jayraj Ugarkar

authorHOUSE®

AuthorHouse™ UK Ltd.
500 Avebury Boulevard
Central Milton Keynes, MK9 2BE
www.authorhouse.co.uk
Phone: 08001974150

First published by AuthorHouse 3/22/2010

ISBN: 978-1-4343-9725-6 (sc)

Library of Congress Control Number: 2008909523

Printed in the United States of America
Bloomington, Indiana

This book is printed on acid-free paper.

For my parents – for being the pillars of my life

For my parents-in-law – for their unflinching support

For my wife – for making my life beautiful

For my son – for putting a smile in my heart

For my sisters – for their love and warmth

Preface

The significance of telecommunications to the development of humankind is undeniable. No other industry sector can claim to touch the lives of more than half the world's population so often, and be such an integral part of day-to-day life.

The initial role of telecommunications was to instantaneously connect people and reduce the spatial distance. However, over the years, telecommunications has shaped human lives in a manner unlike any other technological revolution – or for that matter any other industry. It has seamlessly connected different parts of the world, increased awareness amongst societies and significantly changed the lives of many people. It has enabled expertise to be available when and where required. It has become the top source for news and information, and it plays a major role in bringing entertainment to millions of people.

The economic impact of telecommunications technology is tremendous as well. By the end of 2007, the market was worth more than US$3 trillion, growing at a compounded annual growth rate (CAGR) of more than 6%. This mammoth turnover has created large number of job opportunities, not only in telecommunications but also in many other sectors. According to the International Telecommunication Union (ITU), every dollar invested in telecommunications generates four times as much revenue in other sectors. This statistic alone demonstrates just how important the telecommunications industry is in driving the economy.

It is easy to quantify the economic impact of telecommunications using numbers, but the unquantifiable social impact of telecommunications on the fabric of human society is equally tremendous. It is hard to put numbers on the value of being able to call home from a far-flung place and hear your baby's voice. The Internet has become the main source of information from places where oppressive governments have successfully shut down other modes of communication. Mobile phone, just within a decade and half of its existence, has completely changed the way people connect with each other and has significantly aided growth of business in all sectors. Telecommunications has also decentralized economic activities away from overpopulated urban areas, slowing the exodus from rural areas and creating more employment there.

Human life is increasingly becoming information-based, and telecommunication technologies will continue to play a key role in the rapid evolution of humankind.

Why This Book?

Up until the late 1990s, telecommunication was about networks. However, in the 21st century, it is all about what the customer wants - it is all about services. There is no other industry sector going through such a large-scale change in its underlying core business model. The industry has moved from being network centric with few basic services like voice, TV, and broadband, to a fast paced, vibrant customer centric industry with tens of new services launched every month. The old setup with separate industries for voice, video, and data is shattered. In its place, is a brand new industry providing converged services spanning across communication, media, and entertainment (CME) businesses. An explosion of new technologies like IP Multimedia Subsystem (IMS), Next Generation Networks (NGN), WiMax, 3G/4G are supporting this trend. New services like IPTV, Mobile TV, VoIP, and Web 2.0 based Value Added Services (VAS) are redefining the possibilities. However, the academic world and many books on telecommunications are not addressing these trends. Most of books and courses are either too technical or theoretical, with little relevance to the realities on the ground. The objective of this book is to bridge this gap of knowledge that is required to work in the industry.

The telecommunications industry has always fascinated me, and I have been involved with it since my childhood. I often accompanied my father who worked in a telephone exchange and listened to endless hours of conversation related to every aspect of the business. I studied Communications Engineering and have now been involved in the industry for more than a decade. In spite of all this, I still struggle to get a grip on rapidly changing technology and market dynamics. To get a handle, I started compiling my own notes that explained the basics and the market trends. My focus was not on complex theoretical topics like Fourier's transforms, antennas, or network theory, but on understanding the basic concepts of the various facets of the industry. The focus was on understanding services, which is the main driver behind the rapid changes in market dynamics, and the impact of technological advances. The focus was on understanding trends that were shaping the industry for years to come. Halfway through this compilation activity, I realized that there were many people struggling like me, especially the new ones joining the industry. Even many experienced people only seemed to know about their area of work and had difficulty understanding how what they do affects the larger context of the business. This book is written to address the essentials required to understand and work in the industry.

This book begins with an overview of how the industry started, the contribution of the pioneers to the industry, and the fascinating stories about the invention of services ranging from telegraphy to the Internet. It then moves on to explain the basic terminology and the basic components that are required to support each of the services and applications. It also explains how the industry works, not only on a daily basis by explaining operational activities like fulfillment, assurance and billing processes, but also explains the processes around defining strategies for services, infrastructure and products. It also explains at a high level the agenda of various standards bodies and the impact of regulations on the industry. The final section explains the trends in the market place, covering topics like convergence, next generation networks (NGN), and value-added services (VAS), which are fast changing the very foundation of the business.

Who Should Read This Book?

This book was not written with any particular cross-section of people in mind. Telecommunication is such a diverse and ever-changing field that it is impossible to know every aspect of the industry. The objective of this book is to explain every topic in a simple and lucid manner so that the reader, regardless of whether he or she is a student, newcomer, or expert, can achieve a thorough understanding of the basics. In addition, the language and tone of this book is conversational rather than theoretical. This makes it easy to grasp the crux of the topic easily.

Acknowledgments

Writing this book has been an exercise in endurance. Even though it required tons of patience and hard work from my side, I do not think I could have done it without help from those around me.

I want to start by thanking all my bosses so far: Abhay Juvekar (Syntel), Gary Hirata (First Hawaiian Bank), Alon Rodin (Amdocs), Mary (Amdocs), Harry Cliffe (ATT), Robert Johnson (Infosys), Thomas Viviano (Infosys), Jeremy Kloubec (Infosys), and Mandeep Singh Kwatra (Infosys). I would not have been where I am today without your support and encouragement. Each one of you not only helped me grow and aim higher, but also taught me many valuable lessons in life.

I want to thank my seniors at work: Atul Kahate (Syntel), Thomas Brinker (ATT), Denis Bagsby (ATT), Bill Winkler (ATT), Michael Eyles (BT), Deepak Swamy (Infosys) and Sanjoy Paul (Infosys). The majority of what I know today is because of you. Thanks for being such great teachers and role models.

I want to thank all my colleagues at work, especially Jim Suda (ATT), Linda Lahue (ATT), Denise Kennet (ATT), Chris Strieter (ATT), Stephen Gevers (ATT), Lori Carter (ATT), Abud Claudio (ATT), Maruthi Nori (ATT), Jason Hunt (IBM), Chris Dalrymple (IBM), Geoff Allen (IBM), Rohit Sampige (Infosys), Anurag Johri (Infosys), Vijoy Gopalakrishnan (Infosys), Amy Brooks (BT), Debi Hackett (BT), and Timothy Bridgland (BT). It is not only a pleasure working with all of you, but also an honor to be associated with the best brains in the business.

I want to thank all my teachers. Your tireless efforts and commitment have empowered me to achieve beyond my capabilities.

I want to thank all my friends. It is a pleasure to know each one of you, and I consider your company to be the greatest gift of my life.

I want to thank the production team at Author House for working so hard to realize this book. Thanks to Tim Clarke from Wordsru.com for doing a superb copy-editing job.

Finally, I want to thank my family and in-laws. I am the luckiest person on earth to be surrounded by so many loving, caring, and supportive people.

Jayraj Ugarkar
June 20th, 2008
London, UK

Companion Website

A companion website for this book has been setup for the readers and others to get additional and up-to-date essential information about the various aspects and trends of the telecommunications industry. The website address is www.cme-essentials.com. It provides links to host of white papers, case studies, news, and blogs.

Table of Contents

Section 1 – Overview of the Telecommunications Business

1. Overview of Telecommunications

Telecommunications is defined as any transmission, emission or reception of signs, signals, text, images, videos, sounds or intelligence of any nature by wire, radio, optical or other electromagnetic systems. In other words, it encompasses any communication over distance through telephone, television, radio, network, or other means. The industry encompasses voice, video (TV), and data services provided over wires, over the air (wireless), and via cable and satellite.

The focus of this book is mainly on two-way communication systems like wireline, wireless, cable and satellite. As such, this book does not address one-way communication systems like broadcast TV (analog/DTT) and radio.

Telecommunications is an extremely fascinating and fast-paced industry that affects every aspect of human life. The importance of telecommunication services is undeniable. It has created unprecedented opportunities for all the people of the world to be informed, entertained, and be connected with each other. It enables the coordination of activities between spatially separated people, bridges cultural gaps, enables resources to be available when and where they are needed, and contributes significantly to improved efficiency in virtually every industry.

The telecommunications industry has experienced enormous growth around the world during the last decade because of technological developments and deregulation policies. It now touches more lives around the world than any other services industry. Today, the world has over 2.9 billion mobile connections, 1.85 billion fixed lines, and more than 1 billion Internet and Cable/Satellite TV users.

On the other hand, the telecommunications industry has also seen radical shifts and upheavals. The combined forces of deregulation, technology shifts, evolving customer needs and preferences, and the convergence of services have primarily driven these changes. The shift from wireline to wireless, circuit-switched to packet-switched networks, narrow band to broadband, and the impending dominance of IP technology are all having a profound impact upon the structure and contours of this industry. In addition, the boundaries between telecommunications, media and entertainment services are blurring, thereby forcing telecommunication companies to rethink their strategies. The entire telecommunications business model is changing from providers offering a single service, such as a wireline phone company providing just a basic voice service, to one of collaboration between different players in order to provide a comprehensive range of services. For example, today's ATT, is a conglomerate formed from a merger between SBC, Bell South (wireline, broadband, IPTV), Cingular Wireless, Echostar (satellite TV), and old ATT (VoIP and long distance).

1.1. Impacts of Telecommunications

During the last decade, the telecommunications industry has grown faster than the overall economy in almost all countries around the world, and represents a significant share of GDP. The annual worldwide turnover from the telecommunications business was US$3.2 trillion in 2007, and is estimated to be worth over US$4 trillion by 2010. However, the real impact of telecommunications is that it has transformed the way individuals, businesses and other parts of society work, communicate and interact. Different macro-economic and firm-level studies confirm higher productivity gains where good telecommunications infrastructure exists. Besides increasing productivity, telecommunications is also transforming economic relationships and processes in both the private and public sectors. Positive impacts in various areas have been observed and measured across both developed and developing countries. Just as e-commerce and tele-working allows companies to reduce costs and increase revenues, e-government made available through telecommunications infrastructure has the potential to save money, increase efficiency, and raise transparency in the public sector. The economic impacts of telecommunications are enormous. Studies conducted by the UN's International Telecommunication Union (ITU) have demonstrated that there is a close relationship between telecommunications and economic development in every country. Another report by ITU/OECD (Organization for Economic Cooperation and Development) stated that investments in telecommunication facilities generate reciprocal investment in trade, industry and agriculture at a rate that averages four times the level of investment in telecommunications itself. This factor is even more significant in developing countries, especially in Asia, Latin America and Africa, where telecommunication facilities are rapidly growing.

The social impact of telecommunications on the developing world is perhaps even more dramatic than in advanced nations. The provision of the first telephone in a village or a community has a much greater multiplier effect on economic development than the investment cost of providing that line. Globalization and outsourcing enabled through telecommunications has helped spread economic benefits to different parts of the world. This has tremendously benefited billions of people in their struggle to earn livelihood. Internet and other means of communication are not only the main source of information during catastrophic natural disasters like tsunamis and earth quakes, but also have successfully channeled much needed help to the needy.

The focus of this chapter is on understanding what telecommunication services are, how they are provided, and by whom. The last section of this chapter peeks into the state of the telecommunications market around the world.

1.2. Services

Services are what telecommunication companies or service providers offer to consumers. They can be broadly classified into the following categories:

- Voice
- Video
- Data or broadband

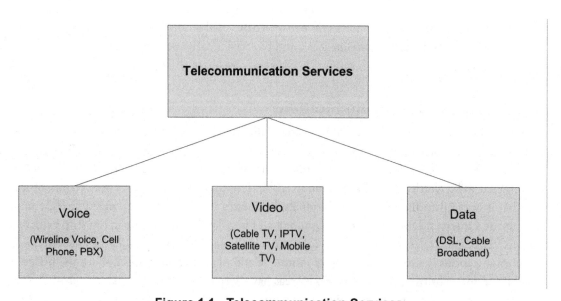

Figure 1.1 - Telecommunication Services

1.2.1. Voice

Voice is the oldest and most commonly provided service. It is provided by wireline, wireless, cable and satellite companies. The voice service provided to consumers can be classified into the following categories:

- Wireline Voice – provided by wireline-based service providers
- Wireless Voice – provided by cellular and satellite-based service providers
- Voice over IP (VoIP) – provided by wireline and cable-based service providers using IP technology

1.2.2. Video

Video/TV service involves the communication of video/TV signals from one central location to many subscribers. Video services these days are provided by wireline, wireless, cable and satellite providers. The video service provided to consumers can be classified into the following categories:

- IPTV – provided by wireline-based service providers
- Cable TV – provided by cable-based service providers
- Satellite TV – provided by satellite-based services
- Mobile TV – provided by cellular-based service providers

1.2.3. Data

Data services refers to the communication of information between machines like PC, laptop, cell phone or PDA. Data services are also provided by wireline, wireless, cable and satellite companies. The data service provided to consumers can be classified into the following categories:

- DSL – broadband (over 128 Kbps) provided by wireline-based service providers.
- Dial-up (less than 56 Kbps) – provided by wireline-based service providers.
- Cable broadband – broadband (over 128 Kbps) provided by cable-based service providers.
- Satellite broadband – broadband (over 128 Kbps) provided by satellite-based service providers.

1.3. Applications

In addition to basic services like voice, broadband or cable TV, service providers also supply applications that run on top of services. These applications provide additional functionality and enhance user experience. Typical examples of applications are:

- Voice mail
- Caller ID
- Call waiting
- SMS/MMS
- Games (e.g., Sudoku, Chess) over mobile and TV

In the industry today, the term "service" is widely used to include applications. The way to differentiate is that an application cannot run by itself, whereas a service can. For example, call waiting is not available on its own, it is available only as part of a voice service.

1.4. Service Providers

Organizations responsible for providing telecommunication services (voice, video and data) are called service providers. They form the largest sector in the telecommunications industry and account for more than 77% of overall industry turnover.

The market place is filled with a variety of service providers. They can be broadly classified based on the transport mechanism used to provide the services:

- Wireline carriers (using wires)
- Wireless carriers (using radio waves)
- Cable company (using hybrid fiber-coaxial cables)
- Satellite communication (using satellites)

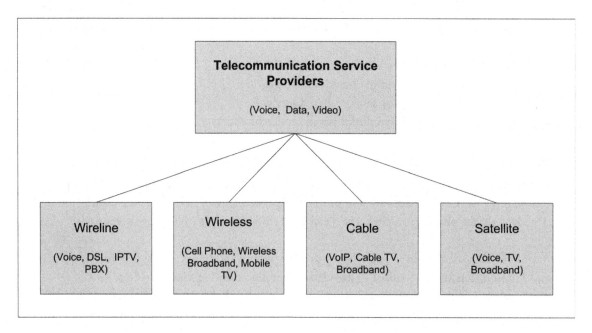

Figure 1.2 - Types of Service Providers

For years, due to regulatory restrictions and technology limitations, service providers provided only one type of service to one type of customer. For example, a wireline service provider only provided voice service to residential consumers. However, things have dramatically changed in the last decade and regulatory authorities no longer restrict companies to serve just one segment (residential or enterprises) or provide just one type of service (voice or video). In addition, IP technology allows a service provider to provide any type of service using the same infrastructure. For example, earlier a wireline provider could only provide voice service to residential customers. However, now due to new technology and changes in regulation, a wireline service provider is allowed to provide broadband and TV services as well. Similarly, wireless, cable and satellite service providers are now providing all three services (voice, video and data). It will not be long before there will be no differentiation between wireline, wireless, cable, and satellite service providers. Each will be capable of providing all the communication and entertainment services required by consumers.

1.4.1. Wireline Carriers

Wireline carriers mainly provide telephone and broadband service via wires that connect customers' premises to central offices maintained by telecommunication companies. The wireline carriers were the first telecommunication service providers and have been in business ever since the invention of the telephone. For years, they primarily provided voice communication services using analog technology over copper wires. Today, the copper wires are being replaced by high capacity optic fiber, and packet based digital technology is replacing analog technology. These two new breakthroughs make efficient use of resources and provide endless possibilities for new services. Packet communication makes it possible to transmit all types of communication signals over the same channel, thereby considerably reducing costs and the price of services.

On the residential side, even though wireline carriers ruled the communication world for over 100 years, their business is in steady decline due to the replacement of wireline telephones with either cell phones or VoIP-based phones. The wireline market in the US is declining from $240 billion (in 2007) at a rate of 2% per annum, while the European market is declining from $104 billion (in 2007) at a rate of 3% per annum. The only success story is broadband DSL and that has been the savior for most wireline carriers in mature markets for quite some time. A successful strategy followed by wireline carriers in order to survive is to provide TV, video on demand and other interactive services by using either DSL or running an optical fiber all the way to the premises (FTTP) or at least to the node (FTTN).

On the enterprise side, the market for wireline-based data connections continues to show steady growth. Wireline service providers provide broadband, IP-based services like IP Centrex, voice PBX, and data networks like LAN, MAN and WAN based on IP, MPLS, ATM and frame relay. In addition, they provide managed services like Web server farms, data centers, VPNs, and network security.

The network built by the world's wireline carriers is referred to as the public service telephone network (PSTN). PSTN connects central offices around the world using many switches interconnected by coaxial or fiber cables. PSTN is still mostly analog, but many parts of it are being converted to digital, especially the connections between central offices.

1.4.2. Wireless Carriers

Wireless operators, many of whom are subsidiaries of wireline carriers, transmit voice, graphics, data, and provide Internet access over a network of radio towers. The signals are transmitted to and from the end-user device (e.g., mobile phone, pager) through the nearest antenna. The antennas are in turn connected to a central office through a wireline network. The biggest advantage of wireless communication is that the infrastructure required to start business can be installed very quickly, even over hostile terrain or a heavily populated area. Wireless devices are popular with customers who need to communicate as they travel, residents of areas with an inadequate wireline service, and those who simply desire the convenience of communicating on the move. Increasing numbers of consumers are choosing to replace their home landlines with wireless phones.

Wireless carriers are primarily divided into two main categories:

- GSM-based service providers
- CDMA-based service providers

GSM and CDMA are two competing technologies. GSM is an open source technology started in

Europe with around 73% of worldwide market share, while CDMA is a proprietary technology led by Qualcomm in the United States. Each of these technologies has many more sub-technologies to provide voice and data services. For example, GSM has GPRS and EDGE technology, while CDMA has a series of CDMA2000 technology for data communications. The technology in each of these categories is continuously evolving, with each evolution referred to as a generation. Most providers around the world are using 2.5 or 3rd generation technology today (2008). Each generation has features better than the previous generation, especially around data communication speeds. Wireless operators are deploying several new technologies to allow faster data transmission (more bandwidth) and better handsets, which will make them competitive with wireline carriers. With these developments, wireless carriers can support advanced applications (e.g., games) and mobile commerce activities like selling music, videos and other exclusive content that can be downloaded and played on wireless phones.

Wireless carriers are developing the next generation of technologies that will surpass the current 2.5G and 3G with even faster data transmission. Several new technologies like 4G, HSDPA, WCDMA, CDMA 2000 and WiMax are either being rolled out or are in the works. The replacement of wires with wireless services should become increasingly common because advances in wireless systems will provide data transmission speeds comparable to broadband landline systems. Wireless carriers also provide high-speed localized broadband availability zones or LANs called Wi-Fi (wireless fidelity). Wi-Fi has gained acceptance in many environments as an alternative to a wired LAN. Almost every home in many countries is Wi-Fi enabled these days. Many airports, hotels and other public places offer access to Wi-Fi networks for people with laptops to log onto the Internet and receive emails on the move. These locations are known as hotspots, spread across many cities around the world. By the end of 2007 there were close to 150,000 Wi-Fi hotspots in over 135 countries around the world. They are located in convenient public locations and provide high-speed connection to the Internet. Wi-Fi is fast, reliable, affordable, and secure. Users can surf the Web and connect to their office network using VPN. Wi-Fi comes in three flavors, commonly known as 802.11b, 802.11a and 802.11g. Each of these standards has a varying degree of speed (11–54 Mbps) and distance coverage (up to 30 ft), with 802.11b and 802.11g the most commonly used. Wireless carriers worldwide plan to spend around $12 billion in building up Wi-Fi infrastructure. Some prominent service providers in this segment are Vodafone, O2, Bharti-Airtel, ATT, NTT DoCoMo, Verizon Wireless, Telenor, and China Mobile.

1.4.3. Cable Companies

Cable companies primarily provide live/recorded television services and specialty pay-per-view programs like popular/new release movies or sporting events. Apart from these, they are now competing directly with telephone companies in the all-important voice and broadband business. However, for decades, cable operators had a one-way-only channel from the operator to the customer. The technology and infrastructure in place could not provide effective two-way communication due to signal interference and the limited capacity of conventional cable systems. This meant the customers could not communicate back through that channel making it impossible to provide voice, broadband and interactive video services like video on demand. Fortunately, new technology has reduced signal interference, fiber optic cables and improved data compression techniques have now made two-way telecommunication services like voice, broadband, and video-on-demand possible on cable systems.

Cable companies have a market share of at least 75% or more in many countries worldwide for providing paid television services. For voice services, cable companies are signing up new customers at a very impressive rate thereby providing a significant challenge to wireline voice service providers. For example, in the US by the end of 2007, cable companies had more than 4.7 million voice customers.

It has long been widely recognized that the cable companies are the best poised to pull off the fabled "triple play" and study after study is confirming that they're making the most headway in persuading customers to buy bundles of voice (VoIP), broadband and TV.

Some prominent service providers in this segment are Comcast, Charter, Time Warner, Cox, Foxtel, and NTL.

1.4.4. Satellite Carriers

The two most popular commercial applications for communication involving satellites are direct broadcast satellite (DBS) and maritime communication services. In many markets around the world, DBS is the only alternative to cable companies for paid television channels, although IPTV from wireline carriers is fast catching up. DBS operators transmit programming from orbiting satellites to customers' receivers known as mini-dishes. Just like cable companies, a lot of revenue is generated through pay-per-view sports or movies.

DBS providers command a respectable market share of 25% or more in many countries around the world. Many DBS providers also provide Internet access, but this is currently not very popular. Satellites carriers are also involved in maritime communication by providing a full range of communication services to ships at sea. Here, satellite carriers have a 100% share as no other existing technology can cater to this market. Satellites are also extensively used for communication purposes by the defense departments of many countries.

Some of the popular satellite carriers are Sky (UK), DishNetwork(US), DirectTV (US) and Inmarsat (maritime).

1.4.5. Virtual Network Operators

Virtual network operators (VNO) provide a variety of telecommunication services to their customers, just like wireline or wireless service providers, but the key difference is that they do not own a network. Instead, they buy the required network capacity wholesale from network owners and use it to sell their services. Even though VNOs appear to compete directly with the network operators from whom they borrow the network, in reality it is a symbiotic relationship – VNOs siphon off some customers, but in return they deliver important wholesale business to the carriers. In addition, VNOs cater to a very niche market based on lifestyle (music or extreme sports) or demographics which the larger network operators cant cater to effectively. For example, a VNO might support the Asian or Hispanic market by providing TV channels in their language or creating unique international calling plans that cater to its customer base. In effect, VNOs give established operators a steady and relatively risk-free source of revenue. VNOs usually re-brand the service as their own, and have their own sales channel and customer care department.

The concept of VNO is quite popular in the wireless market, where they are referred to as mobile virtual network operators (MVNO). One of the best examples of MVNO is Virgin Mobile UK, an extremely popular brand in the UK closely associated with a trendy and posh image. It operates primarily in the transport and music business and is quite a visible brand across the country. Virgin has also cashed in on this brand image and started Virgin Mobile, which is now a very profitable and successful mobile service provider. It has a very innovative pricing scheme, some value-added services, and a good customer care department. Virgin managed to be twice voted "Best Network Operator," while its host operator was voted "Worst Mobile Operator," even though they both were running off the same network.

Some of the popular VNOs are Disney mobile, ESPN mobile, Virgin mobile, Xtreme mobile.

1.5. Service Channels

Telecommunication service providers use different mechanisms to provide services to customers. Service channels describe what physical medium is used to transport services. There are two main channels:

- Wireline (copper, fiber, cable)
- Wireless (radio waves, satellite)

1.5.1. Wireline

As the name suggests, services provided through this channel are via some type of wire (copper wire, cable, fiber optic).

This is the oldest, but the slowest growing segment in telecommunications today. Regardless of the slow growth, wireline voice service continues to be the most important source of revenue and margins for the bulk of service providers around the world. It is true that in most mature markets like North America and Europe, many people are disconnecting their wireline telephone and are switching over to mobile or VoIP, but there is still robust demand in emerging markets.

The savior of wireline operators has been DSL based broadband. It is estimated that there are more than 300 million broadband users today, many of whom use DSL. As the years pass, much of the activity among current wireline operators will be focused upon providing high bandwidth connectivity to homes and enterprises through advanced versions of DSL and fiber. It is expected that such high bandwidth connections will be required to support next-generation services like IPTV (internet protocol TV), plug and play devices (e.g., gaming consoles), and of course the Internet. At the enterprise level too, broadband connectivity will be needed extensively for Video Conferencing, WLAN (wireless local area network), IP Centrex, and VoIP services.

1.5.2. Wireless

Services provided through this channel do not use wires to transmit information, but instead use radio waves.

Mobile wireless services are the fastest growing segment in telecommunications today, with over 2.7 billion mobile phone users in 2007. Affordable, reliable and ubiquitous wireless services have profoundly altered the way people access the global communications network. The principal growth driver will continue to be mobile telephony, with more than 80% of the revenue coming from voice. Price competition is driving the margins down, but overall demand will continue to keep pace, especially from developing countries.

There are two main technologies for wireless services today – GSM and CDMA. The capabilities of these technologies are standardized as generations (e.g., 2G, 3G). The main technology generation supporting today's wireless networks are 2G and 2.5G networks, which are good for voice communications but not capable of providing good bandwidth. The future of wireless lies in providing high bandwidth services (e.g., mobile TV, GPS based services) and attractive Web 2.0 applications (e.g., games, social networking apps like Facebook or Orkut). This will be possible with the next generation of wireless technologies like 3G and 4G.

1.6. Service Offerings

The next step in understanding the telecommunications marketplace is to understand the types of services offered to residential and enterprise customers. As mentioned earlier, there are hundreds of types of services offered to these two market segments. Some services like voice for residential are easy to comprehend as everybody uses them, but the same voice service to enterprises involves hundreds of combinations, and can be very difficult to understand. At high level, the service offerings can be classified as follows:

* Residential offerings
* Enterprise offerings (business, universities, government organizations)

1.6.1. Residential

The communication requirements of residential consumers are not very complicated. It usually involves one or two voice lines, connectivity to the Internet, and a television connection.
* Voice Services

Voice services to residential customers are provided by wireline, wireless and cable providers, but each uses different techniques and technologies. Wireline service providers typically provide voice service using analog technology over copper wires, or use IP technology (VoIP) over copper or fiber. Wireless service providers use digital technology and transmit using radio waves. Cable providers use IP technology (VoIP) and transmit using cable.
* Data Services

Data services involve the transportation of information from machine to machine like desktop PC, laptop, cell phone or PDA. Typical individual consumers at home use data services for connecting to the Internet. Data services are also provided by wireline (DSL), wireless (EDGE, HSDPA or EVDO), cable (DOCSIS) and satellite service providers.
* Video/Television Services

Video/television services involve the communication of video/TV signals from one central location to many subscribers. Examples of consumer video services include television and video on demand (VOD). Unlike voice and data, video services to residential customers are mainly provided by cable and satellite companies. However, wireline and wireless companies are fast catching up by providing these services using IP technology.

1.6.2. Enterprise

The enterprise market comprises small, media and large businesses, universities and government organizations. Unlike residential customers, the communication requirements of enterprises are varied. The requirements of small and medium-sized enterprises are very different from those of large enterprises with global operations. The range of products, features and solutions available for enterprise in each service type (voice, video and data) is quite large.
* Voice Services

Voice services used by enterprises differ significantly in terms of features from the ones used by residential users. Typical enterprise establishments require anywhere from a few to a couple of thousand voice terminals. In addition, some require advanced features like hunting, direct inward/

outward calling, multiple call appearances, line restriction, call monitoring, meet me lines, priority ringing, and a host of other features that are typically not found in residential consumer voice services. Voice service is provided to enterprises in two ways:

o   Premises-based solutions (PBX – private branch exchange and IP-based PBX)
o   Central office-based solutions (Centrex – central office exchange service and IP-based Centrex)

In premises-based solutions (PBX), most of the equipment required, like switches and administrative consoles, is all located at the enterprise's premises, while in Centrex all the equipment is located at the service provider's central office.

•   Data Services –

Data services provided by services providers to enterprises are of two types:

o   Bandwidth (DS0, T1, OC)
o   Network (LAN, MAN and WAN)

The bandwidth requirements for enterprises are quite varied as compared to residential users. Small enterprises are usually content with just a connection to the Internet, but large enterprises require a dedicated high-speed link between their geographically dispersed enterprise locations. To satisfy these varied requirements, the bandwidth provided ranges from simple DS0 (64Kbps) to T1 line (1.54 Mbps) to OC-192 (10 Gbps). A DS0 is capable of supporting one voice line and provide Internet access. A T1 line has a capacity of 1.544 Mbps and can carry a maximum of 24 voice communication lines and provide Internet access. Most small enterprises prefer a T1 line as it provides enough voice lines and a connection to the Internet. Medium-sized companies with higher requirements go for a T3 line, which has a capacity of 43 Mbps (28 T1s). Large enterprises with thousands of employees, and big office buildings spread across continents go for OC1 (51.84 Mbps) to OC192 (10 Gbps). With this level of capacity, they can connect all their geographically dispersed offices and get phones, computing resources, Internet connectivity, and video conferencing infrastructure.

Service providers also help enterprises network (or connect) their communication infrastructure. The network can cover a single location (LAN – local area network) or connect multiple LANs in a city to form a MAN (metro area network) or connect geographically dispersed office through WAN (wide area network).

•   Video Conferencing

A real-time video session between two or more users or between two or more locations is referred to as video conferencing. It has become increasingly popular with large enterprises that have high bandwidth dedicated links between their offices as it saves time and money.

1.7.   Other Players

In addition to service providers, many other players in the telecommunications industry actively support the service providers in running their business:

•   Hardware manufacturers

- Software manufacturers
- Professional consultants

1.7.1. Hardware Manufacturers

Hardware companies manufacture equipments required to build and support the telecommunications business. Following are some examples:

- Switches and routers – Physical devices that route network traffic from source to destination
- Modems – Convert signals from one format to another, which helps disparate networks and devices talk to each other
- Cables – Wires that connect communication equipment, includes optical fiber, copper wire, coaxial cable, etc
- Handsets – Equipment used for communication purposes, including mobile handsets and telephones
- Transmitters – Electronic devices that propagate an electromagnetic signal with the aid of an antenna, such as radio (for mobile phones), television, or other telecommunications services
- Servers – Computer servers are extensively used by providers to process usage data, maintain accounts, etc

Most of the equipment manufacturers were hit hard during the dotcom bust, but since early 2004, things have started looking better. There has been a good jump in investment by telecommunication operators, which increased by more than 10% in 2007 to almost US$200 billion worldwide. Mobile operators are at the root of this growth, with 15% to 20% increase in their investments every year. Geographically, growth has mainly come from emerging economies in Latin America, Eastern Europe, the Middle East, and the Asia Pacific region.

Enterprises are also increasing investment in their telecommunications infrastructure, most notably for migrating from PBXs (private branch exchange) to IP (Internet protocol) compatible solutions, setting up of WLAN infrastructures, and general upgrade of infrastructure.

Some prominent hardware manufacturing companies are Cisco, Nokia, Ericsson, Lucent-Alcatel, Corning, Siemens, Motorola, and Huawei.

1.7.2. Software Manufacturers

Software manufacturers provide the software required to build and support the telecommunication business. Examples of softwares are as follows:

- Network software – A lot of networking equipments, like routers and switches, are now armed with highly advanced software that not only helps perform many functions at much higher speeds, but also provide performance metrics, failure alarms, and have self-healing capabilities
- OSS (operations support systems) – The softwares in this category are used to run the day-to-day operations like customer care, ordering, billing, provisioning, assurance, inventory, and element management
- Next-generation applications – A host of companies today are involved in developing innovative applications like find me–follow me, commerce, games, presence-based, location-based and productivity improvement-based applications

- End-user device software – This category includes vendors who supply operating systems for end-user devices like mobile handsets, set-top boxes, and modems
- Middleware platforms – The next generation of services like IPTV (Internet protocol-based TV) and VoIP (voice over IP) require extensive use of software capable of supporting and delivering the service
- Convergence enabling software– Convergence is mainly aided by softwares like SDP (service delivery platform) and IMS (IP multimedia subsystem)
- Operating systems and databases – Telecommunication companies are some of the largest users of all types of servers like mainframe, midrange, and PCs, and have some of the largest databases in the world. A significant portion of the IT budget in telecommunication companies is spent in managing these server farms and data centers
- Enterprise management – Telecommunication companies extensively use software to run the company itself. Good examples are the software used to support accounting, payroll, HR, and intranet

Many telecommunication companies around the world have now stepped up their investment in software to improve customer service, support the rapid growth in demand, and reduce expenses. The worldwide market for OSS alone is growing at 7% CAGR, and stood at over $40.5 billion in 2007. Overall software spending has grown at an even faster rate of 11% CAGR to over $100 billion.

Some prominent software manufacturers include Alcatel-Lucent, IBM, BEA, SAP, Oracle, Amdocs, Telecordia, Nokia, Ericsson, and Microsoft.

1.7.3. Professional Consulting Service Providers

Professional consulting service providers provide a variety of services to support telecommunication businesses in areas like IT, risk assessment, financial audit, strategy, and others. The worldwide telecommunication consulting market was worth an estimated US$11 billion in 2007, representing a growth of 10% compared with 2006.

Some prominent consulting service providers are Accenture, IBM, Infosys, TCS, EDS, Deloitte Consulting, Gartner, and Ovum.

1.8. Worldwide Markets

The telecommunications market around the world is quite varied. In some regions, it is mature and saturated; while in others, it is growing like wildfire. The worldwide telecommunications market can be categorized into the following regions:

- Asia
- Latin America
- Africa
- North America, Europe, and Australia

1.8.1. Asia

Asia is the fastest growing region, with overall growth upwards of 9.3% during the last five years.

The Asian telecommunications market was estimated to be worth around US$300 billion in 2007. The big drivers are wireless, broadband and IP services, particularly as value-added services come into the market. Fuelling this robust growth are China, India and Indonesia, Philippines, which collectively contributed to about 80% of the net subscriber additions in 2007. In spite of constant price wars and shorter product lifecycles, cellular subscription in the Asian region is expected to continue achieving double-digit growth in the next four to five years.

The broadband situation is not much different. Asia is the world's largest Internet market with an estimated 375 million Internet users. In the area of broadband Internet access, South Korea continues to be a world leader with over 70% of households having a broadband connection.

In spite of this tremendous growth, it is important to note that subscriber penetration of cellular phones is still under 20%, even in countries like India. The subscriber penetration for broadband is even less, currently standing at around 10%. These figures and the fact that Asia is home to half the world's population clearly shows that this region will remain the hotbed of growth for the next decade or so.

1.8.2. Latin America

This is another fast-growing region, dominated by Mexico, Venezuela and Brazil. Many of the countries in the Latin American region combine a rapidly expanding middle class with the increasing privatization of key industries. Broadband grew at an annual rate of around 54% in 2007, making Latin America one of the world's fastest growing regions in terms of broadband uptake. However, broadband penetration at the end of 2007 was only 2.5%, considerably less than the global average of 5.4%. In addition, mobile telephony is one of the most dynamic industries in Latin America. By early 2007, mobile penetration in the region had surpassed the 50% milestone, but there are considerable variations from country to country.

1.8.3. Africa

This is another region exhibiting robust growth. However, most of the growth in this region is from the mobile market as the lack of wireline infrastructure makes mobile telephony the only alternative. Just like the Asian and Latin American markets, the African market is also characterized by fiercely competitive prices, low cost handsets, very innovative pre-paid schemes.

1.8.4. North America, Western Europe and Australia

These regions are considered as the most mature telecommunications market with almost 100% of their population having access to at least basic communication services. The market is not growing as fast as in other regions of the world, but still generates a large proportion of worldwide revenue. In addition, the North American and Western European markets are the hub of innovation. They are at the forefront of trials in next-generation technologies such as IPTV, 3G, FTTH, WiMax, NGN, and value-added services.

2. Telecommunications Revolutions

2.1. The Seven Revolutions

As fascinating and complex as the telecommunications industry today is, the history of telecommunications is even more interesting. This chapter is not only meant to introduce the pioneers who have significantly contributed to the growth of the telecommunications industry over the years, but also to shed light on some of their stories that depict how it all happened.

The growth of telecommunications since the invention of telegraphy during the early 1800s can be divided into seven big revolutions:

1. Telegraphy (started in 1830)
2. Telephony (started in 1880)
3. Cable (started in 1948)
4. Networks (started in 1960)
5. Satellite (started in 1963)
6. Wireless (started in 1979)
7. Internet protocol (IP) (started in 1983)

2.1. Telegraphy

Telegraphy, invented around the 1830s, was the earliest system invented for the purpose of telecommunications. Surprisingly, it is still used very widely around the world within the military, airline and financial industries. Telegraphy uses Morse code to send messages that are often referred to as telegrams or cable/wire messages. Telegrams are still widely used in many developing countries for sending messages when the destination is not easily contactable through telephone or other means of modern communication. They are sent by a network of tele-printers called telex machines, usually located in post offices. The message is then delivered by a postal worker to the addressee.

2.2.1. The Inventor

The earliest device for telegraphy was invented by Samuel Morse in 1837. The device relied upon the electromagnet that had been invented only a few years earlier. It used a series of dots and dashes to represent letters and numbers, called Morse code.

2.2.2. The Story

In 1835, Samuel Morse was appointed to teach painting at New York University, but he was very interested in magnetism and electricity. During a trip to the United States from Europe on a ship, he overheard a conversation about electromagnetism that inspired his idea for an electric telegraph. Even though he had little training in electricity, he realized that pulses of electrical current could convey information over wires. After months of work and help from his colleague Leonard Gail, Morse developed a rudimentary code to represent alphabets and numbers and an apparatus to transmit them. Finally, in 1837, a successful experiment was carried out using 1,700 feet of copper wire coiled around the room. The experiment proved that signals could be transmitted by wire. Pulses of current functionally equivalent to dots and dashes were transmitted, which deflected an electromagnet marker reproducing the codes on a strip of paper. The code then had to be decoded into letters and numbers using a dictionary composed by Morse. He gave a public demonstration of the device and the code in 1838, but it was not until five years later, in 1843, after a lot of persuasion from Morse, that Congress funded $30,000 to construct an experimental telegraph line from Washington to Baltimore, a distance of 40 miles.

Samuel Morse and his partners immediately commenced construction of the 40-mile line between Baltimore and Washington. By May 1844, the line was completed and on May 24, 1844, Samuel Morse sat before his instrument in the room of the Supreme Court at Washington and flashed the message "WHAT HATH GOD WROUGHT!" to his partner Alfred Vail 40 miles away in Baltimore. Vail flashed back the same momentous words.

After this auspicious start, telegraphy slowly started gaining popularity. Ezra Cornell, founder of Cornell University, took up the task of laying lines across the country, connecting city with city. In the meantime, Samuel Morse and Alfred Vail improved the details of the mechanism and perfected the code. Many more pioneers came after them and added further improvements. Telegraphy is still extensively used to send personal messages and transmit news around the world by the military, merchant ships, and the financial and airline industries. It became a rapid success and gave birth to modern electronic data communications.

2.3. Telephony

Telephony (wireline-based) is the oldest and still the most popular means for conducting voice communication today. The first commercial telephone exchange, with just 21 subscribers, was operational in 1878 in New Haven, CT, USA. By 1911 there were 6 million subscribers in the United States alone, with many millions more around the world. Telephony, without a doubt, has played a major role in shaping human lives.

2.3.1. The Inventors

Although Alexander Graham Bell is widely credited in the history books with the invention of the telephone, the history of the telephone is mired with claims from many different inventors, including Antonio Meucci and Elisha Gray. It is fascinating to read the stories of these men and realize that Bell was not really the one who invented the telephone.

Alexander Graham Bell
A pioneer in the field of telecommunications, Alexander Graham Bell was born in 1847 in Edinburgh, Scotland. He moved to Ontario, Canada and then to the United States, settling in Boston before beginning his career as an inventor. In 1876, at the age of 29, Alexander Graham Bell invented the telephone and filed for a patent at the US Patent Office on March 7, 1876. The patent covered "the method of, and apparatus for, transmitting vocal or other sounds telegraphically." By a strange coincidence, another great American inventor, Mr. Elisha Gray, applied just two hours later on the same day for patent of a similar kind of device.

Apart from being a great inventor, Bell was also an entrepreneur. Bell and others formed the Bell Telephone Company in July 1877. In 1879, it merged with the New England Telephone Company to form the National Bell Telephone Company. In 1880, they formed the American Bell Telephone Company, and in 1885 the American Telephone and Telegraph Company (AT&T), which in 1899 became the overall holding company for all the Bell ventures.

Elisha Gray

Prior to filing a patent for telephone on march 7th, 1876, Elisha Gray, on February 14, 1876, filed a caveat with the US Patent Office describing apparatus for transmitting vocal sounds telegraphically. A caveat is a confidential, formal declaration by an inventor stating his intention to file a patent on an idea yet to be perfected. Caveats were filed as a means of protecting an idea from being usurped by fellow inventors. Claims and counterclaims were followed by litigation to determine who the real inventor was. Ultimately, based on its earlier filing time by a mere few hours, the US Patent Office awarded Bell, not Gray, the patent for the telephone.

Antonio Meucci

Italian inventor Antonio Meucci claimed to have invented a telephone-like device called a **teletrophone** much earlier than Bell in 1849 while in Havana, Cuba. He based his experiments on the goal of transmitting speech by electric current. Antonio moved to Staten Island, USA and continued his research. He demonstrated his invention in 1860 and had a description of it published in New York's Italian language newspaper. He eventually filed for a caveat with the US Patent Office in 1871, but unfortunately, due to serious burns on a ferry accident, lack of English knowledge and poor business abilities, Meucci failed to develop his inventions commercially in America. After his fervent claims, US Congress recognized Meucci's work, and in 1887 moved to annul the patent issued to Bell on the grounds of fraud and misrepresentation, a case that the Supreme Court found viable and remanded for trial. However, Meucci died in October 1889, the Bell patent expired in January 1893, and the case was discontinued without ever reaching the underlying issue of the true inventor of the telephone. However, the 107th US Congress re-opened the issue, and by passing resolution 269 on September 25, 2001, it officially credited the invention of the telephone to Antonio Meucci instead of Alexander Graham Bell.

With so many claims and counterclaims, the question of who invented the telephone remains debatable. A note left by Elisha read, in part, "The history of the telephone will never be fully written.... it is partly hidden away ... and partly lying on the hearts and consciences of a few whose lips are sealed – some in death and others by a golden clasp whose grip is even tighter."

The important thing to note here is that there is no single inventor of the telephone. Many people contributed to the telephone as it is today, and each is worthy of recognition for their contribution. The invention of the telephone has changed the world dramatically, and it is extremely difficult to imagine how life would be without phones.

2.3.2. The Story

Of all the inventors and early contributors to the telephone, Alexander Bell's work is the best documented. Bell worked as professor of Vocal Physiology at the University of Boston and trained teachers in the art of instructing deaf mutes how to speak. One day in 1874, while conducting an experiment on a make-break circuit driven by a vibrating metal strip, he accidentally found that it could be vibrated by disturbing another metal strip that was within the influence of the same magnetic field. This discovery led him to believe that all the complex vibrations of speech might be converted into sympathetic currents, which in turn would reproduce the speech at a distance.

Bell and his assistant Watson discovered that the movements of the metal strip alone in a magnetic field could transmit the modulations of the sound. Bell devised a receiver consisting of a stretched diaphragm made of goldbeater's skin and an armature of magnetized iron attached to its middle but free to vibrate in front of the pole of an electromagnet in circuit with the line. This apparatus was completed on June 2, 1875, and on the same day, he succeeded in transmitting sounds and audible signals. On July 1, 1875, he instructed his assistants to make a similar membrane receiver, which could be connected to the first by a wire. He kept one piece in a room and the other in the cellar of his home in Boston. Bell, held one instrument in his hands, while Watson listened at the other in the cellar. The inventor spoke into his instrument, "Do you understand what I say?" and the delighted Mr. Watson rushed back upstairs and answered, 'Yes!" This was the first one-way telephone call. However, the first successful two-way telephone call was not made until March 10, 1876 when Bell spoke the famous words, "Mr. Watson, come here, I want to see you," into the device and Watson answered. This was the day that today's telephone was truly born.

2.4.   Cable

Cable was first used in villages across the US in 1948 as a means to get paid television channels. Since then, the industry has grown into a thriving and global telecommunications force. Today, the cable communications industry has 75% market share, not just in the US, but also around the world in the paid television channels business. The cable industry has now expanded beyond providing just paid television channels to become one of the largest broadband providers, and is expanding into providing voice communication services in the form of voice over IP (VoIP).

The total number of subscribers using cable around the world is close to 350 million, with around 80 million subscribers in the United States alone. However, the fastest growing markets are India and China, accounting for close to 60% of all new cable subscribers during 2007.

2.4.1. The Inventors

Just like the telephone, there are many claimants for the title "inventor of cable." Ed Parsons of Astoria, John Walson of Mahanoy City, Martin Malarkey, Jr. of Pottsville, and Robert Tarlton of Lansford, are among those credited with building early CATV (community antenna television) systems in 1948.

2.4.2. The Story

Although cable television systems are now present in every corner of the world, it began out of necessity. Consumers in many rural areas across America were unable to receive conventional television broadcast signals. Amateur geniuses figured out that the solution was to capture and retransmit the signals. This was achieved by using extremely high antenna towers or microwave repeater stations. These antennas and towers intercepted over-the-air signals and retransmitted them to the households that could not receive them using regular VHF or UHF antennas. Such retransmission systems spread across remote and rural America throughout the 1950s and '60s. By 1960, there were 640 companies with 650,000 subscribers, and by 1970, these numbers had grown to 2,490 companies with 4,500,000 subscribers. Most of these companies were generally "Mom-and-Pop" operations with few channels. Since those early days, the industry has grown into a thriving and global telecommunications force, with many large corporations operating in multiple communities. Worldwide subscription of cable television continues to grow rapidly, and the cable industry has not only expanded into providing multiple channels of sports, entertainment, news and public affairs, but also voice and broadband connections.

2.5. Networks

The concept of linking two or more computing devices together to exchange data and share resources is referred to as networking. Networks are formed by connecting machines like desktop computers, servers, printers, storages devices, etc. The earliest form of data communication was telegraphy, but widespread machine-to-machine communication did not take off until businesses began using computers for data processing in the 1960s. Up until 1960, data was passed between computers using tapes, and then the ARPANET (Advanced Research Projects Agency Network) team showed the world how data could be exchanged between computers using wires. By the late 1960s, data communication between computers using telephone lines became commonplace in businesses throughout the world. Universities were the first to set up vast networks of computers located within and outside the campus, and large corporations followed suit.

Residential consumers did not use networking concepts until the Internet was introduced in 1983. At around the same time, the price of personal computers started falling to affordable levels, so they were commonly found in households in many countries. The Internet fueled the growth of networks, consumers started using the Internet as a source of information, and businesses grew through business-to-consumer commerce (B2C) and business-to-business (B2B) commerce.

Networks are thriving today because of two key components:

- Industry standard communication protocols like Ethernet, SONET, ATM, Frame Relay and MPLS
- Supporting infrastructure like Optical Fiber and Routers

2.5.1. The Inventors

Ethernet

Ethernet is the core technology that has made LAN so popular. It was first conceived and developed by Robert Metcalfe and David Boggs (Metcalfe's assistant) during 1973. However, it took them another three years before they could publish their work in a paper entitled "Ethernet: Distributed Packet-Switching for Local Computer Networks" in 1976.

SONET

SONET (synchronous optical network) is a standard for connecting fiber-optic transmission systems across the world. SONET was proposed by Bellcore in the mid-1980s and is now an ANSI standard. The rapid growth of SONET and other higher layer protocols like ATM, Ethernet, and IP running over SONET has resulted in the realization of new services like video conferencing and video on demand, as well as providing more bandwidth for the increasing volume of traditional data.

Optical Fiber

Optical fiber is a medium through which information containing modulated light can be transmitted from source to destination. Light has been used for centuries as a means of communication, but it was not until the invention of the laser in the 1960s that optical fiber was conceived as a medium for communication. Optical fiber has become the de-facto medium for building networks and is very commonly used to build high bandwidth networks.

Routers

A router is a central switching device in a network that directs and controls the flow of data through the network. It is usually built as a combination of hardware and software. It is highly capable of making very complex and intelligent routing decisions that keeps the data traffic on a network flowing. It finds the best path for a data packet to be sent from one network to another by first determining all possible paths to the destination address and then picking the most expedient route, based on the traffic load and the number of hops.

2.5.2. The Story

Ethernet

Robert Metcalfe, the inventor of Ethernet protocol, left Xerox in 1979 to start a company called 3Com to promote the use of personal computers and local area networks (LANs). He successfully convinced Digital Equipment, Intel, and Xerox Corporations to work together to promote Ethernet as a standard. By March 1981, 3Com shipped its first Ethernet hardware to the public. In 1983, the IEEE published the Ethernet standard 802.3. Xerox turned over all its Ethernet patents to the nonprofit IEEE, which in turn licensed any company to build Ethernet hardware for a fee of $1,000. This made Ethernet the most widely used LAN technology. In 1989, the Ethernet standard won international approval with the decision of the International Standards Organization (ISO) to adopt it as standard. Now an international computer industry standard, Ethernet is used in about 90% of installed LANs around the world.

SONET

SONET was originally designed for the public telephone network. In the early 1980s, the forced breakup of AT&T in the United States created numerous regional telephone companies, and these companies quickly encountered difficulties in networking with each other. Fiber-optic cabling already prevailed for long-distance voice traffic transmissions, but the existing networks proved unnecessarily expensive to build and difficult to extend long-haul data and/or video traffic. The American National Standards Institute (ANSI) successfully devised SONET as the new standard for these applications. Like Ethernet, SONET provides a "layer 1" or interface layer technology (also termed physical layer in the OSI model – chapter xy). As such, SONET acts as a carrier of multiple higher-level application protocols like asynchronous transfer mode (ATM), frame relay and Internet protocol (IP) packets.

The key difference between Ethernet and SONET is the distance over which these protocols are effective. SONET is effective over cabling that spans distances of thousands of miles, while Ethernet can only go up to 300–600 feet.

Optical Fiber

There are many significant contributors to the development of optical fiber, but the optical fiber as you see it today was first conceived by Robert Maurer, Donald Keck and Peter Schultz in 1970. They began experimenting with fused silica, a material capable of extreme purity with a high melting point and a low refractive index. A single strand of fiber optic wire was capable of carrying 65,000 times more information than a strand of similar sized copper wire. However, the very large loss of light in the optical fibers prevented coaxial cables from being replaced. Early fibers had a signal strength loss of around 1,000 dB/km, but this was not good enough for transporting communication signals. Research showed that the loss had to be reduced to at least 20 dB/km for any practical use.

In 1972, Corning Glass Works made the first fiber with a loss of just 4 dB/km. Since then, technology has progressed rapidly and the fiber currently in use today has a signal strength loss of less than 0.5 dB/km.

Another factor that was limiting early implementation was the cost and weight of the laser equipment used to generate the light for producing the signals. Fortunately, continued development in semiconductor technology led to smaller sources of light like the LED (light emitting diode). LEDs, first introduced in 1971 by Bell Laboratories, are exclusively used as the source of light in fiber optical communication today.

All these advances in fiber optics have led to almost exclusive use of fiber optics in long-haul systems. Today, more than 80% of the world's long-distance traffic is carried over optical fiber cables. More than 25 million kilometers of the cable that Maurer, Keck and Schultz designed has been installed worldwide.

Routers

The router was first defined by an international group of computer networking researchers called the International Network Working Group (INWG). It was set up in 1972 as an informal group to consider the technical issues involved in connecting different networks. The definition was explored and realized by DARPA and Xerox. The first IP router was developed by Virginia Strazisar as part of the DARPA-initiated effort, during 1975-1976. The first multiprotocol routers were created by staff researchers at MIT and Stanford in 1981; the Stanford router was done by William Yeager, and the MIT one by Noel Chiappa.

2.6. Satellite

A communications satellite is a satellite stationed in space for the purpose of telecommunication. They are predominantly used by defense departments, merchant shipping organizations, airlines, satellite navigation or GPS, and for the direct broadcast of television signals to individual households. The advantage of satellites is their ability to carry simultaneously many diverse communication applications (data, voice and video) to practically any place on earth. It is used to establish basic communication services quickly in countries where there is no telecommunications infrastructure in place.

The satellite services sector more than tripled in size between 1996 and 2007, with revenues passing the $100 billion mark in 2004. Satellite service providers integrate trans-oceanic terrestrial networks, broadcast and two-way voice, video and data services to commercial, government and military consumers.

2.6.1. The Inventors

American Telephone and Telegraph (AT&T) was the first company to recognize the potential of using satellites for communication purposes. With help from NASA, AT&T launched the Telstar satellite, that was placed in a low orbit to serve anyone with an antenna as it passed over them. However, this was not very practical, as it did not provide a continuous communication link. A solution was first provided by Harold Rosen from Hughes Aircraft, who proposed a satellite that would operate in geostationary orbit. Based on this proposal, NASA launched the first geostationary satellite, Syncom 2, in 1963. Since then, groups of geostationary satellites have become the mainstay of satellite communication.

2.6.2. The Story

The United States also took the lead in creating the organizational framework for satellites. In the early 1960s, the Comsat Corporation was created when President John F. Kennedy signed the Communications Satellite Act and paved the way for the formation of the first commercial enterprise dedicated to satellite communications. In 1963, Comsat began operating and established its headquarters in Washington DC. During August 1964, Intelstat (International Telecommunication Satellite Consortium) was formed at the proposal of the United States in order to develop a global network involving telecommunication agencies from around the world. Today, Intelsat owns 51 communication satellites and covers 99% of the world's population. It is the world's largest commercial satellite-based provider, serving leading media and broadcasting organizations, multinational corporations, and telecommunication companies.

2.7. Wireless

Wireless communication is the fastest growing sector in telecommunications. "Wireless" is an old-fashioned term for a radio transceiver, but now the term is used to describe modern wireless connections such as those in various cellular networks, Wi-Fi, wireless LAN, WiMax, and other wireless broadband networks.

The wireless revolution has swept the world with such alacrity that in just two decades the number of wireless phones is greater than the number of wireline phones. According to the International Telecommunications Union (ITU), current wireline subscriptions around the world stand at 1.85 billion, while wireless subscriptions top 2.7 billion.

The concept of wireless communication is nothing new, but the wireless revolution took off after the introduction of the concept of a cell (explained in detail in chapter 7). The first cellular-based wireless telephone service became operational in Japan in 1979, and since then the wireless industry has achieved exponential growth worldwide. The global wireless business is now worth more than US$450 billion, and growing very rapidly. China is the world's largest wireless cell phone market with over 260 million subscribers, with the US second with 153 million. India is the world's fastest growing market with a compounded annual growth rate (CAGR) of at least 20% per year.

The wireless data market is showing equally phenomenal growth, led by 2.5G, 3G and Wi-Fi (802.11x) technologies. Wireless hardware and software used in data communications has shown a staggering growth rate of over 200% in recent years, which is expected to continue. By the end of 2007, over 50% of households and over 30% of companies in North America and Europe had already wireless enabled their LANs. The number of worldwide Wi-Fi hotspots grew by 87% to 100,355 hotspots in 115 countries.

The future for wireless is very bright, with continued explosive growth expected from voice, video and data markets around the world. 3G and WiMax are touted to be the next-generation technology for wireless broadband services, which will eventually make a wireless "triple play" containing voice, video and data possible. There is considerable interest in wireless VoIP technology by enterprises around the world, which promises to be a significant next step.

2.7.1. The Inventors

Bell Labs introduced the idea of cellular communications in 1947 with police car technology. However, Dr Martin Cooper, a general manager for the systems division at Motorola, is considered the inventor of the first modern portable handset.

2.7.2. The Story

Cooper always believed that given a choice, people in the future would demand the freedom to communicate without being restricted to a location. He made the first call on a portable cell phone in April 1973, to his rival, Joel Engel, the head of Bell Labs Research. By 1978, public trials of a prototype cellular system were conducted by AT&T and Bell Labs. In 1979, the first commercial cellular telephone system began operation in Tokyo. In 1981, Motorola and American Radio Telephone started a second cellular radio telephony system test in the Washington/Baltimore area. By 1982, FCC finally authorized commercial cellular service in the USA. The first American commercial analog cellular service or AMPS (advanced mobile phone service) was made available in Chicago by Ameritech in 1983. The same year, Motorola first introduced the 16-ounce "DynaTAC" phone into commercial service, but the exorbitant cost of $3,500 for each device slowed the process of commercial acceptance. It took another seven years before there were a million subscribers in the United States in 1990.

Despite the incredible demand, it took cellular phone service more than three decades to become commercially available in the United States. However, today more people use cell phones than wireline phones around the world. Cellular phones have also evolved considerably, now weighing as little as three ounces, and cellular service providers not only provide voice communications services, but a host of other services like entertainment, Internet, email, photo, and video.

2.8. The Internet

The Internet is used by just about everyone, but very few people realize that it is simply a network of computers linked together. Every time you go online you are just adding one more computer to a network formed by million of users like you, businesses, and other organizations. In a sense, it is the largest network in the world, growing at a rapid rate. Imagine, on average, the content available on the Internet doubles every 100 days.

Thus, the Internet can be defined as a worldwide, publicly accessible network of interconnected computer networks that transmit data by packet switching using the standard Internet protocol (IP). It is a "network of networks" that consists of millions of smaller domestic, academic, business and government networks. These interconnected networks together support various services, like news, e-commerce, e-government, entertainment, email, chat, file transfer, and many more.

One question that is often asked is "Who owns the Internet?" The answer is "no one!" Since it is a network of multiple computers, it exists because all these computers want to connect to each other. If for any reason all these computer owners decide not to connect together, there will be no Internet. As users logging on to the Internet, we are all contributing to its existence, popularity and growth. However, there is a authority called ICANN (Internet Corporation for Assigned Names and Numbers) that coordinates the assignment of unique identifiers on the Internet, including domain names, Internet protocol (IP) addresses, and protocol port and parameter numbers. A globally unified namespace is essential for the Internet to function so that there is only one holder of each name. ICANN is overseen by an international board of directors drawn from across the Internet's technical, business, academic, and non-commercial communities.

A key point to note here is that there is no governing body for the Internet as it is a distributed network comprising many voluntarily interconnected networks. Anybody can join the network and exchange information.

Contrary to common usage, the Internet and the World Wide Web or Web are not the same. The Internet is a collection of interconnected computer networks, while the Web is a collection of interconnected documents, linked by hyperlinks and URLs. The Web is accessible via the Internet, along with many other services including email, file sharing, chat, voice, bit torrent, and others.

One of the key protocols that has made the Internet feasible and so popular is TCP/IP (transport control protocol/Internet protocol). TCP/IP is a set of communication protocols upon which the Internet and most commercial networks run. The popularity of the Internet is not just based on it being a vast network of interconnected computers allowing people to exchange information and communicate. It is because of TCP/IP protocol, which makes it very easy to communicate and for applications like the World Wide Web, Email, and tools like Web browsers to run.

2.8.1. The Inventors

The Internet
In 1962, J. C. R. Licklider of MIT wrote papers that first envisioned a globally interconnected set of computers through which everyone could quickly access data and programs from any site. In the same year, Licklider became the head of DARPA (Defense Advanced Research Projects Agency), the first US federal government-funded computer research program. Leonard Kleinrock of MIT/UCLA also contributed significantly by developing the theory of packet switching, which forms the basis of Internet connections. By 1966/7, research on connecting different computers and exchanging information between them had developed sufficiently, leading to the creation of a computer network system called ARPANET. The major contributors to the design of ARPANET were teams at MIT, the National Physics Laboratory (UK), and RAND Corporation, executed by BBN of Cambridge, MA under Bob Kahn. ARPANET was first tested in October 1969 by establishing communication between computers installed in UCLA and Stanford. The tests were successful, and this heralded the birth of the Internet.

TCP/IP
TCP/IP protocols were first outlined in a 1974 paper by Bob Kahn and Vincent Cerf. It was introduced in 1977 for cross-network connections, and it slowly began to replace earlier protocol called NCP (network control protocol) within the original ARPANET. TCP/IP was faster, easier to use, and less expensive to implement.

Both Bob Kahn and Vincent Cerf are recipients of the prestigious Presidential Medal of Freedom. They also received the Turing Award, which is the equivalent of the Nobel Prize in the computing industry.
Email
In March 1972, Ray Tomlinson at BBN wrote simple software that was capable of sending and receiving messages, motivated by the ARPANET developers' need for an easy coordination mechanism. Roberts later expanded its utility by writing the first email utility program that could selectively list, read, file, forward, and respond to messages. From there, email took off to become one of the most popular uses of Internet.

World Wide Web
The World Wide Web only began in March 1989 at CERN (originally named after its founding body, the Conseil Europeen pours la Recherché Nucleaire, it is now called the European Laboratory

for Particle Physics). Tim Berners-Lee made the proposal to create a simple scheme to incorporate several different servers of machine-stored information already available at CERN. This scheme was to use hypertext to provide a single user-interface to many large classes of stored information such as reports, notes, databases, computer documentation, and online systems help. This was the birth of the Web as we know it today, where information and services are available from one user to another.

Web Browser

The first widely used Web browser was Mosaic, version 1.0 of which was released in 1993 by Marc Andreessen and Eric Bina from the University of Illinois at Urbana-Champaign. They renamed it Netscape Communicator for commercial launch.

Microsoft launched its popular Web browser Internet Explorer in 1995, which quickly became very popular. Then in 2002, an open source version of Netscape called Mozilla was released. Mozilla is quite a popular browser capable of running on non-Windows platforms like UNIX and Linux, and is available free as part of an open source initiative.

These and many more visionaries have contributed significantly to the development of the Internet and its applications since inception.

2.8.2. The Story

The Eisenhower administration reacted to the USSR's 1957 launch of Sputnik with the formation of the Advanced Research Projects Agency in 1958, in an attempt to regain a technological lead. After much work from Licklider, Kleinrock, and many others, the first node with interconnected computers went live at UCLA on October 29, 1969; this network became known as ARPANET. By 1971, ARPANET had 15 nodes connecting 23 computers, and by 1973, connections to the UK and Norway were established. Following this, the British Post Office, Western Union International, and Tymnet collaborated to create the first international packet-switched network, referred to as the international packet-switched service (IPSS), in 1978. This network grew from Europe and the US to cover Canada, Hong Kong and Australia by 1981.

TCP/IP was initially designed to meet the data communication needs of the US defense department. It began as a partnership between ARPANET, US universities and the corporate research community to design open, standard protocols and build multi-vendor networks. In 1974, the design for a new set of core protocols for ARPANET was proposed in a paper by Vinton G. Cerf and Robert E. Kahn. The official name for the set of protocols was TCP/IP Internet Protocol Suite, commonly referred to as TCP/IP, which is taken from the names of the network layer protocol (Internet protocol – IP) and one of the transport layer protocols (transmission control protocol – TCP). TCP/IP is a set of network standards that specify the details of how computers communicate, as well as a set of conventions for interconnecting networks and routing traffic. The initial specification went through four early versions, culminating in version 4 in 1979.

The first TCP/IP wide area network was operational by January 1, 1983, when the United States' National Science Foundation (NSF) constructed a university network backbone that would later become NSFNet. This date is considered to be the birth of the Internet. This was followed by the opening of the network to commercial interests in 1985. Use of the term "Internet" to describe a single global TCP/IP network also originated around this time.

The network gained a public face in the 1990s. On August 6, 1991, CERN, which straddles the border between France and Switzerland, publicized the new World Wide Web project two years after Tim Berners-Lee had begun creating HTML, HTTP and the first few Web pages at CERN.

In 1993, the National Center for Supercomputing Applications (NCSA) at the University of Illinois at Urbana-Champaign released version 1.0 of Mosaic, which became a very popular and easy to use tool for "surfing" or "browsing" the Internet. By late 1994, there was growing public interest in the previously academic/technical Internet. In 1995, Microsoft hosted an Internet Strategy Day and announced its commitment to add Internet capabilities to the popular Windows 95 and all its other products. In fulfillment of that announcement, Microsoft Internet Explorer arrived as both a graphical Web browser and the name for a set of technologies.

Meanwhile, over the course of the 1990s, the Internet successfully accommodated the majority of previously existing public computer networks. Its organic growth is often attributed to the lack of central administration, as well as the non-proprietary open nature of Internet protocols, which encourages vendor interoperability and prevents any one company from exerting too much control over the network.

The Internet is growing at an annual rate of 18% and now has over a billion users. At this rate, the next billion users will follow within the next 10 years.

2.8. Telecommunications Timeline

The following timeline of the telecommunications industry provides a very interesting insight into how the industry grew through a variety of fronts over the years.

1840–1870

In 1840, Samuel Morse patented the practical telegraph, and in 1844 set up a 40-mile telegraph line between Washington DC and Baltimore.

1870–1890

On March 10, 1876, the telephone was born when Alexander Graham Bell called to his assistant, "Mr Watson! Come here! I want you!" By 1878, the first commercial telephone exchange opened in New Haven, CT with 21 subscribers. By 1881 the first commercially successful long-distance line of 45 miles between Boston and Providence, Rhode Island was opened for business. AT&T was incorporated in New York in 1885.

1891–1910

Up until 1891, all the calls made had to be connected by an operator. The caller had to call the operator and they would then make a manual connection by placing a connector end into the desired connection and signaling the location being called by cranking and manually ringing the phone. The problem here was that once connected, the operator had to listen to the conversation to detect the end of the phone call and manually disconnect it. Needless to say, the operators were among the best informed in the community! When the nuisance of operators knowing everything going on in a community became annoying, Almon Strowger began working on a way around it in 1888. Finally, in 1891 he came up with a central office switching system whereby the telephone user was not dependent upon the operators. After the telephone, the switch was the greatest invention, which continues to change the landscape of telecommunications today.

1911–1930

By 1911, there were more than 6 million telephone lines in the United States. On January 25 1915, the first transcontinental line from New York to San Francisco opened. Alexander Graham Bell died at his summer home in Nova Scotia in 1922, and in his honor, the service of the entire telephone system of the United States and Canada was suspended for one minute during the funeral service.

In 1924, the first transmission of pictures over telephone wires was publicly demonstrated by Bell engineers. A public demonstration of television by wire from Washington DC to Bell Telephone Laboratories in New York City was made in 1927. The first color photographs were sent over wire from San Francisco to New York, for the *New York World*. In the same year, the first transatlantic call was made. In 1929, a telephone was installed on President Hoover's desk. Up until this time, the President talked from a booth outside his executive office!. On June 27 the same year, the first public demonstration of color television was made at Bell Laboratories in New York.

1931–1950

In 1934, the Communications Act became effective in the USA. Approved by President Roosevelt, it brought interstate telephone business under regulation by the Federal Communications Commission (FCC). In 1938, the first crossbar central office installation went into service at Troy Avenue, Brooklyn. The nationwide numbering plan, as we have it today, came into effect in 1946. Mobile telephone service opened along the Boston–Washington highway in 1947.

1951–1970

In 1955, the first transatlantic telephone cable was laid between Newfoundland and Scotland. Russia launched the first satellite in 1957 (Sputnik01), and the US followed suit in 1958 by launching the first satellite for telecommunication. Customer trials of the world's first electronic telephone exchange began in Morris, Illinois in 1960. The first international direct dial from London to New York was accomplished in 1970. The same year, the FCC formally announced its plans for regulating the cable television industry. In 1968, DARPA selected BBN to develop ARPANET, the father of the modern Internet.

1971–1990

By 1971, the number of telephones in service under Bell Systems reached 100 million. Independent telephone companies served an additional 25 million in the US, and an estimated 160 million were in use in some 200 other countries. The first digital switch was installed by AT&T in 1976. The FCC authorized cellular trials in 1977. IBM introduced the desktop personal computer (IBM PC) in 1981.

The AT&T divestiture of 1984 resulted in the creation of seven regional Bell operating companies. The US reached its one-millionth cellular subscriber in 1987, the same year that ADSL was introduced in the USA. In 1989, Ethernet won international approval with the decision of the International Standards Organization (ISO) to adopt it as standard number 88023. The first transatlantic fiber optic cable was completed in 1988.

1991 – Present

There have been lots of changes in just about every sector of telecommunications in almost every country on earth. Mergers and acquisitions have become common, state ownership is being reduced, and technology is changing rapidly. Today, there are more cellular phone users (2.7 billion) than wireline users (1.85 billion). More than a billion people use the Internet, and wireless has become ubiquitous. Internet protocol (IP) technology is breaking the product silos in the market, enabling cellular, wireline, cable and satellite companies to provide the entire gamut of telecommunication and entertainment services. New means of voice communication like VoIP and Wi-Fi are fast replacing traditional wireline telephones. Broadcast TV transmission is being seriously challenged by highly interactive IP-based TV, and the speed of broadband connection, whether wireless or wireline is increasing day by day. Overall, the last 15 years have seen more changes in telecommunications industry than in its entire history.

Finally, ever wondered how every telephonic conversation starts with a "Hello"?
Apparently Alexander Graham Bell suggested "Ahoy, ahoy", but fortunately most telephone users were not pirates. The word "Hello" was first recommended by Thomas Edison as a way to greet someone when answering the telephone and it became widely accepted. Now you know that the electric light bulb was not Edison's main contribution to the world!

Section 2 – Basic Concepts of Telecommunications

3. Basic Concepts of Telecommunication Technology

The telecommunications industry is quite complicated as it has many service providers providing a variety of services using many different technologies. However, even with all this complication and diversity, some of the basic terminology and technological concepts dealt with in this chapter form the foundation over which the entire industry is built.

3.1. Basic Communication Systems

Let us start with what a basic telecommunication system looks like. A telecommunication system is responsible for the transmission of a signal from a sender to a receiver. At its simplest (Fig. 3.1), the system contains the following components:

- Signal source – The source of a signal may be a human being or a machine, and the signal may be simple voice, text, video, or a combination of all three. This signal is fed into a source device
- Coder – A coder receives the signal and converts it into a suitable format that can be transmitted
- Transmitter – A transmitter transmits the coded signal using a transmission medium like copper wire, fiber, or radio waves
- Channel – A transmission channel acts as the physical link between the sender and the receiver
- Receiver – A receiver receives the signal from the transmission channel
- Decoder – A decoder decodes the signal into a format suitable for the receiver to comprehend the information
- Playback – A receiving end device plays the signal back in its intended format

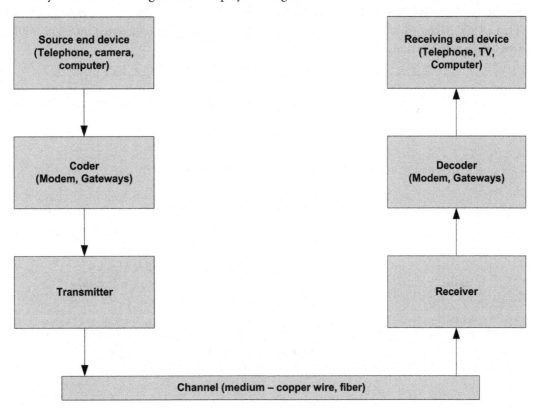

Figure 3.1 - Basic Components Of Telecommunication Systems

3.2. Transmission

This section explains the terminology used for transmitting information from source to destination.

3.2.1. Signal

A signal is a means of passing information from source to destination. In the world of telecommunications, a signal is a form of energy (electricity or light) passed through a medium like copper wire, optical fiber or airwaves to reach its destination. There are two types of signals used in telecommunication systems:

- Analog
- Digital

3.2.2. Frequency

Signals can be in the form of light or an electrical signal. They travel as waves (if analog) or a stream of bits (if digital), as shown in Fig. 3.2 (representative only, actual waves differ), with each wave carrying information. Once generated, each type of wave travels at a constant speed, but the number of waves generated by the source per second differs. Frequency is defined as the number of waves emitted by a source per second, and is measured in hertz. One hertz (Hz) is one complete wave per second, one kilohertz (kHz) is one thousand waves per second, one megahertz (MHz) is one million waves per second, and one gigahertz (GHz) is one billion waves per second. Human beings are capable of generating and hearing waves in the range of 20 Hz to 20 KHz, but the range of frequencies generated by telecommunication devices is quite large, ranging from a few Hz to about 1 petahertz (Phz = 10^{15} hertz). For example, your cell phone (GSM) operates in the range of 850 MHz to 1900 MHz, which means that every second of your speech is converted into 850 million waves before they are transmitted.

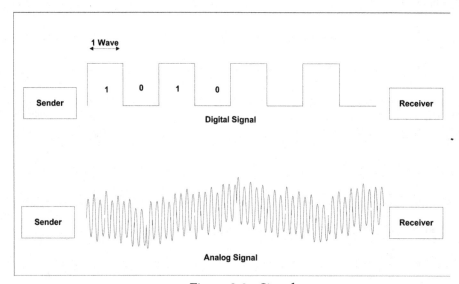

Figure 3.2 - Signal

3.2.3. Noise

Noise is an extraneous signal that mixes with the source signal and distorts it. Noise is often caused by interference from nearby wires, magnetic fields (electric motors), from the equipment used, and sometimes introduced by the medium itself.

3.2.4. Electromagnetic Spectrum

The possible range of frequencies is referred to as the electromagnetic spectrum. The range of frequencies available is infinite, but the frequencies used in telecommunications lie within 300 KHz to 1 PHz. Frequencies below this range are just not practical for telecommunication purposes, and higher frequencies fall into the x-ray and gamma ray range, which are harmful, highly susceptible to interference, and do not travel very far before requiring amplification. Fig. 3.3 depicts the range of frequencies and the applications associated with them. The range of frequencies selected for an application depends upon various factors. The general theory is that lower range frequencies are capable of traveling longer distances than higher range frequencies before requiring amplification, but are not capable of serving as many customers. Other influential factors are the medium used for transmission and its availability.

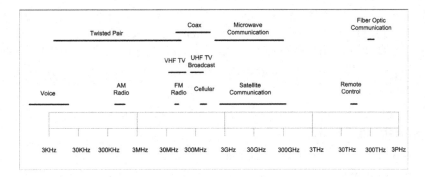

Figure 3.3 - Electromagnetic Spectrum

3.2.5. Bits and Bytes

In traditional telecommunication systems, signals are made up of waves, as shown above in Fig 3.2. However, in telecommunication systems where computers are used, signals are made up of bits. Unlike humans, who use alphabets, numbers and special characters for communication, computers use sets of 8 bits (or 16, 32, 64, 128) to represent alphabets, numbers and special characters. For example, in the ASCII coding standard, the number 9 is represented by 01010111, and the letter A is represented by 01100101.

3.2.6. Bandwidth

In digital communication, bandwidth is a measure of how fast information can be carried. It is defined as the maximum number of bits that a medium can transmit per second. The higher the bandwidth, the faster it can carry information. For real-time communication (like voice and live TV), higher bandwidth also means better quality. For example, a TV channel requires 2 Mbps of bandwidth for good quality viewing. If the bandwidth of the channel is less than 2 Mbps, the picture will not be clear. For non-real-time communications, like music downloading or reading a newspaper online, lower bandwidth only means slower speeds; the quality will remain the same as the source. For example, a line with a bandwidth of 2.5 Mbps can download a 5 Mb

MP3 song in 2 seconds, while on a 1 Mbps line it will take 5 seconds. The quality of the song does not change because it took longer to download or because it was downloaded on a channel with lower bandwidth.

In analog communication, bandwidth is a measure of the range of frequencies from the original signal that are successfully transmitted. For example, if the range of frequencies produced by a speaker is between 2 KHz and 20 KHz, the bandwidth required is 18 KHz (20 KHz minus 2 KHz). If the bandwidth of the medium is just 2 KHz then the sound heard at the receiving end will not be very clear at all. Thus, the bandwidth for analog communication is defined as the range of frequencies that a medium can successfully transmit, measured in hertz. The greater the range of frequencies a medium can handle, the greater the clarity. In addition, with multiplexing technique, a channel with higher bandwidth can carry multiple conversations too.

3.2.7. Signaling

Telecommunication systems around the world contain many components like switches, gateways, controllers and servers. Signaling is the process by which nodes communicate to establish, maintain and tear down a communication session between two or more parties.

For voice communication, the signaling between a telephone and the network is called dual tone multi-frequency (DTMF). There are also many new signaling protocols like SS7 for switch-to-switch signaling in PSTN, and H.323 and SIP for VoIP communications. Each of these is explained in detail in the next few chapters.

3.3. Transmission Mode

The transmission mode describes the direction of flow of signals in a communication medium. There are three main types:

3.3.1. Simplex

A simplex transmission supports the transmission of signals in one direction only. Broadcast and cable TV are good examples of simplex transmission. You may not have realized it, but the communication is one way only - you get to watch the shows or hear the radio, but cannot communicate back.

3.3.2. Half Duplex

Half duplex supports the transmission of signals in both directions, but only one direction at a time. Remember how in some old John Wayne movies he used to say "Over" when he was done saying something? This was because the transmission mode was half duplex and "Over" was the signal for the other person to start talking.

3.3.3. Full Duplex

Full duplex supports the transmission of signals in both directions at the same time. Most modern communication systems are full duplex.

3.4. Switches

There are billions of telephone users around the world. Have you ever wondered how a connection between you and your friend living hundreds of miles away is established? The answer is through switches. Groups of switches work in tandem to establish a connection between two users. The term switch or switching system is also used for central offices. There is usually one or more piece of switching equipment per central office, and a typical switch can handle anywhere from 100 to upwards of 10,000 communication lines.

During the early days of telephony, telephone exchanges had manually operated switching equipment in order to connect two telephones. Today's switching equipment is far more advanced. They are not only capable of supporting voice, but also support data and video services.

Through the years, the following types of switches have been used in the telecommunications industry:

- Manual switching
- Electro-mechanical switches
- Digital switches
- Soft switches

3.4.1. Manual Switches

Manual switching was the first switching technique to be designed. It involved an operator (or group of operators for cross-country calls) manually connecting two telephones, as shown in Fig. 3.4. Early telephones did not have a dial pad as we see today; all telephones were connected to an operator manned switchboard located at a central location (central office or telephone exchange). When the user lifted the telephone receiver, they would be connected directly to the local operator. The user would then let the operator know the number of the receiver, and the operator would manually connect the two users. When the call was complete, the operator released the connection. The first manual exchange was installed in New Haven, USA in 1878.

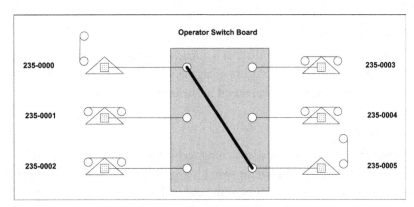

Figure 3.4 - Manual Switching

3.4.2. Electro-mechanical Switches

As the number of telephones increased, manual switching was proving to be inefficient and expensive. The fact that operators listened in to all the conversations also created social problems. After having difficulties with the local telephone operators, Almon Strowger, an undertaker from Kansas City, was motivated to invent an automatic telephone exchange. He believed that a customer's call for an undertaker was routed to his competitors rather than him because the operator did not like him. He was convinced that subscribers should choose who was called for the undertaker services, rather than the operator. He first conceived his invention in 1888, and patented the automatic telephone exchange in 1889.

The first automatic telephone exchange was a combination of electrical and mechanical systems, and it took a couple of milliseconds to operate the relay and switch the calls. The Strowger switch quickly gained popularity, and many more people contributed to the development of the electro-mechanical switch. Electro-mechanical switching systems were initially expensive, unreliable, and required a lot of maintenance, but as they evolved, they became cheaper and more reliable. Carriers were able to reduce labor costs by reducing the number of operators. They were also able to serve more customers.

3.4.3. Digital Switches

Electro-mechanical switches were a remarkable improvement over manual switching, and served the industry for many decades. However, there were quite a few problems with electro-mechanical switches: they were highly inefficient as they had a lot of moving parts, consumed a lot of electricity, and required an army of technicians to maintain them. In addition, they found little use beyond voice transmission. Digital switching equipment or electronic switching system (ESS) was invented to overcome the deficiencies of electro-mechanical switches. They are entirely made up of non-moving parts like transistors, and integrated circuits (ICs).

3.4.4. Soft Switches

This is the latest in switching technology. Even though digital switches are quite advanced, but they are still very expensive as they use a lot of hardware components like transistors and chips. A soft switch performs switching functionality entirely by means of software running on a computer server.

3.5. Switching Techniques

Switching is about connecting a transmitter and a receiver so that communication can take place between the two. Due to the complex nature of signals (voice, video and data) used in transmission, there a variety of switching techniques are used in the public network.

3.5.1. Circuit Switching

Circuit switching (Fig. 3.5) was the first switching technique to be developed, and is still used for voice over wireline telephone. In circuit switching, a dedicated circuit is established between the caller and the receiver at the beginning of the call and held for the entire duration of the call. This technique is very good for voice conversation as there is minimal delay and the audio quality is good too. However, this technique is not very efficient for data communication. A couple of milliseconds of delay during the download of a page on the Internet does not cause any noticeable irritation to the user, but can be very disruptive during a voice communication.

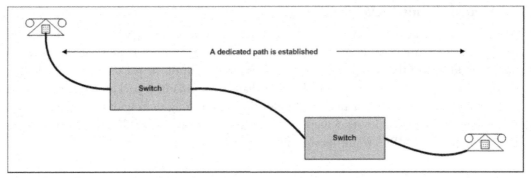

Figure 3.5 - Circuit Switching

3.5.2. Packet Switching

Packet switching (Fig. 3.6) was developed to overcome the inefficiency of circuit switching in data communication. With the packet-switching technique, no dedicated circuit is set up, and the information does not reach the destination in real time. Information at the transmitting end is gathered and broken into equal-sized blocks (usually 1000 bytes) called packets. Each packet contains not only the data to be transmitted, but also the destination address, the sequence number of the packet, and some more control information. Smart computers called routers use complex algorithms to pick a route in the right direction so that the packet can reach its destination. The router also makes sure that the route chosen is not clogged or down. Each packet may take different paths to reach its destination and may arrive out of order. Once the packet reaches its destination, the destination router acknowledges the receipt (called handshaking information) to the sender. The packets are then arranged in the right order and handed over to the receiving application. If any packets are missing, the destination router makes a request to the sending computer to resend those packets.

The efficiency of a packet-switching system is nearly 100% as the channel is used only when required. The biggest drawback of packet switching is that there is a minor delay in the packet reaching its destination, but this is satisfactory for data transmission where a certain time delay is permissible. For years, packet switching was considered unsuitable for voice communication, as even a minor delay can throw off a voice conversation completely. However, packet-switching systems are now tremendously fast and the time it takes for a packet to reach its destination is drastically reduced. VoIP (voice over Internet protocol) is a good example of using the packet-switching technique for voice communication.

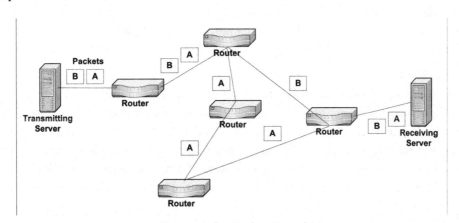

Figure 3.6 - Packet Switching

3.5.3. Virtual Circuit-Packet Switching

Virtual circuit-packet switching makes the best of both worlds. A preplanned route is established from source to destination. Packets only contain the address of the next node on that path. At each node, the address of the next node is given; this is repeated until the packet reaches its destination. Virtual circuit-packet switching is much better than either circuit or packet switching because no dedicated circuit is established and no complex time consuming routing of individual packets is required. Major practical implementations of this type of switching are X.25, ATM (asynchronous transfer mode) and MPLS (multi-protocol label switching).

3.6. Connectivity

The following are some of the terms used to describe the links that help establish connectivity between a transmitter and receiver.

3.6.1. Circuit

A circuit is the bidirectional physical path that runs between two pieces of equipment. For example, the copper wire connection from the telephone at home to the cross connect box in the neighborhood is a circuit.

3.6.2. Line

A line is the bi-directional physical path between any two entities, and contains multiple circuits. For example, the copper wire connecting the telephone at home to the central office is a line. This contains a circuit from home to the cross connect box in the neighborhood, and another circuit from the cross connect box to the central office.

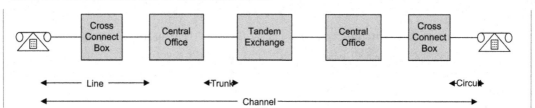

Figure 3.7 - Terminology Reference

3.6.3. Trunk

A trunk is the connection between central offices or telephone exchanges. It is usually a high-capacity transmission medium like optical fiber, microwave, or coaxial cable.

3.6.4. Channel

A channel is the complete bi-directional conversation path between transmitter and receiver, and refers to the band of frequency or time slot used for a single two-way conversation. A channel is established using many circuits, lines and trunks, and describes the whole end-to-end communication path. For example, a wireless call from London to Paris involves a wireless circuit from the cell phone to the nearest tower, another circuit to the central office in London, a trunk line between London to Paris, and another circuit from the central office in Paris to the tower nearest to the called cell phone.

3.7. Transmission Media

Transmission media refers to any material substance like copper wire, coaxial cable, fiber optic, or free space that is used for the propagation of signals from one point to another. The signals can be electrical, light, microwave or radio wave, carrying modulated voice or data.

During the early days of telephony, copper wire was the only medium available for connecting subscribers. Today, there is a variety of media used for telecommunications, ranging from simple copper wire to optical fiber to radio waves. Usually a mix of each of these transmission media is used for completing a connection. For example, central offices may be connected with each other using optical fiber, the connection from central office to the cross connect box may be 50/100 pair copper wire or DS1/DS3 cable, and from cross connect to home is usually twisted pair copper wire. Radio waves are used if wireless devices are used for communication. Each of these transmission media has a certain capacity, cost structure, and quality as shown in Table 3.1.

So, in all, there are four types of media used for transmitting information today:

- Copper wire
- Coaxial cable
- Microwaves
- Fiber

Medium	Capacity/speed	Vulnerability to noise	Cost	Availability
Copper wire (unconditioned and conditioned type)	Low capacity 300 bps to 28 kbps (unconditioned) 64kbps to 1 Mbps (conditioned)	High	Low	Widespread (last mile in wireline connections)
Coaxial (base and broadband type)	High capacity 10 Mbps to 264 Mbps (base-band) 10 Mbps to 550 Mbps (broadband)	Low	Medium	Medium (used extensively by cable companies)
Microwave (terrestrial and satellite)	High capacity 12 Mbps to 50 Mbps (terrestrial) 56 Mbps to 274 Gbps (satellite)	Low	Medium– High	Widespread
Optical Fiber	Up to 30 Gbps	Zero	High	Catching up

Table 3.1 – Comparison of key attributes of different transmission media

3.7.1. Copper Wire

Copper wire or twisted pair copper wire was the first and still very widely used medium used for voice communication. It is mainly used for transmission from cross connect box to the equipment at the customer's workplace or home. There are miles and miles of copper wire laid out, especially as the "last mile" connection in the field across the world today, and providers are forced to reuse them for applications beyond voice. Recent developments in technologies have made such reuse possible in applications like high-speed broadband, IPTV, and video on demand.

3.7.2. Coaxial Cable

Coaxial cable is very widely used by the cable industry to connect from the central office to homes. It has a thick copper wire in the core that carries the signal, and is surrounded by another concentric layer of insulation that runs along the same axis (hence "coaxial"). Due to the nature of construction of the cable, it is capable of carrying very high frequency signals without leakage. Cable companies transmit not only multiple television channels, but also high-speed broadband connection on a single coaxial cable.

3.7.3. Microwaves

Microwaves are high frequency radio waves ranging from 1 GHz to 170 GHz used for point-to-point communication of voice, video and data signals. Microwaves have been used in the telecommunications industry since the 1940s, and their reliability and throughput is proven in the field. They are typically used in the following networks:

- GSM mobile phone networks use the lower microwave frequencies in the range of 1.8 GHz to 1.9 GHz (GSM also uses the 450 MHz, 800 MHz and 900 MHz range).
- Wireless LAN protocols, such as Bluetooth and the IEEE 802.11g and b specifications, use microwaves in the 2.4 GHz range, and 802.11a uses the 5 GHz range.
- WiMax (Worldwide Interoperability for Microwave access) is designed to operate between 2 GHz and 11 GHz. The commercial implementations are in the 2.5 GHz, 3.5 GHz and 5.8 GHz ranges.
- In many countries, microwaves are widely used in long-distance communications, with numerous towers placed around the country.
- Microwaves are extensively used to cover live events using small microwave transmitters placed on vans.

The advantages of microwaves are as follows:

- Cheaper and faster as no cables are required to be laid. Tall towers can serve large areas and communicate with other towers if they are in their line of sight
- Frequently used where cables cannot be laid due to terrain or time limitations
- Unlike fiber, which is restricted to point-to-point communications, microwaves are suitable for mobile communication
- Satellite communications use frequencies in the microwave range. The satellite functions like a microwave repeater tower

3.7.4. Optical Fiber

Optical fiber lines are strands of extremely pure glass that are as thin as human hair but have extremely large bandwidth and can carry signals over long distance without needing amplification. Fiber is also immune to electromagnetic interference that plagues many of the other wired media. It is non-corrodible, does not carry electric signals (instead uses light to transmit the signals), is lightweight and flexible. Needless to say, with so many beneficial properties, fiber has become the most popular transmission medium.

The most attractive feature of fiber is the extremely large bandwidth that a single strand can provide. The latest fiber optic technology from Nortel supports 72 light wavelengths on each fiber strand. Depending on the transmitters attached, each wavelength can supply 2.5 Gbps, 10 Gbps, or 40 Gbps. This means that a single strand at the lowest end can support a bandwidth of 180 Gbps. In practical terms, this would allow between 909,000 and 14.5 million simultaneous HD streams to flow over a single fiber optic cable. Compare this with today's DSL or cable broadband that can provide a bandwidth in the range of 1 Mbps to 8 Mbps.

With this type of bandwidth available, telecommunication companies will be capable of not only providing voice services, but also ultra-high-speed broadband Internet connections, hundreds of TV channels, and even movies on demand, all of which require a large amount of bandwidth.

The bandwidth available in fiber is now in the range of terabits per second. This has been made possible due to technological developments like wave division multiplexing. This allows the network operators to deploy new services that require large bandwidth without laying additional fiber. This efficient reuse of existing network resources is not only driving down costs, but is also making newer services more readily available.

Apart from the many advantages, one of the key differentiators of fiber is that even though it does cost more than copper and coaxial cable on a per-mile basis, a large number of coaxial cables can be readily replaced by a single strand of fiber. Telecommunication companies like ATT and Verizon are rolling out fiber all the way to the cross connect box or node (FTTN – fiber to the node) and to the customer's premises (FTTP – fiber to the premises). The current trend in the telecommunications industry is to have fiber laid all the way to the premises, but the fact is that not all of the last-mile copper can be replaced. There is an enormous amount of copper in the last mile, and it doesn't make economic sense to just throw away that investment overnight.

3.8. Signals

Signals are created using either an electrical or light source. The information from the source is converted into one of the two types of signals before being transmitted over the network.

3.8.1. Analog Signals

The telephone networks built during the early days of telephony were primarily designed to carry voice signals. The transmitter in a telephone handset is a microphone that converts the sound waves into electrical signals. The electrical signals created are directly proportional (analogous) to the sound waves emitted by the person speaking into the telephone handset, hence the signals are analog. Analog signals vary both in frequency and amplitude (Fig. 3.8); the pitch of the voice varies the frequency and the loudness varies the amplitude. The electrical signals are then transmitted over telephone circuits and subsequently converted back to sound waves at the receiving telephone handset by a receiver that is nothing but a speaker. Analog signals are still widely used for voice communication between central office and the telephone at a customer's premises.

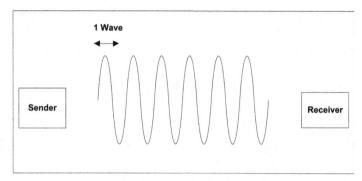

Figure 3.8 - Analog Signals

3.8.2. Digital Signals

Digital signals only came into use in telecommunications systems in 1962. Digital signals were primarily designed for machine-to-machine communications, but lately the human voice is also transmitted (e.g., VoIP). A digital signal comprises streams of 1s and 0s called bits (Fig. 3.9). They are grouped into a set of 8 (called a byte), 16, 32 or 64 bits to represent a number (0–9), a letter of the alphabet (A–Z, a–z), or a symbol (*, /, and others). For example, in 8-bit ASCII, the number 9 is represented by 01010111 and the letter A is represented by 01100101.

Digital signals are extensively used for human voice transmission. Obviously, human beings do not speak ASCII, so to convert the human voice into digital signals, a special technique called pulse code modulation (PCM) is used. The detailed workings of PCM are beyond the scope of this book, but once a voice is converted into a digital signal (1s and 0s), it is transmitted over a digital network just like machine-machine communication.

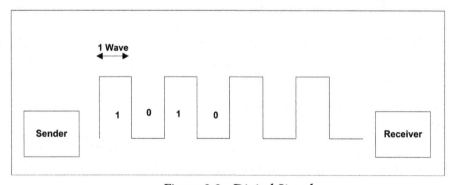

Figure 3.9 - Digital Signal

3.8.3. Comparison

Both analog and digital transmission techniques have advantages and disadvantages. A carrier's choice of using analog or digital network is based upon economics and requirements. In today's telecommunications world, both techniques are extensively used to complete a communication channel.

Criteria	Analog	Digital
Application	Voice transmission only	Voice and data transmission
Bandwidth required	Low, not suitable for high bandwidth applications like video-on-demand	High, suitable for all types of communications
Multiplexing technique	Frequency division multiplexing	Time division multiplexing
Quality of transmission	Low (eliminating noise from signal is difficult)	Very high (eliminating noise from signal is very easy)
Equipment cost	Low	High
Operation cost	High network maintenance cost	Low network maintenance cost
Network intelligence	Analog networks are dumb	Digital networks are intelligent – they issue faults, alarms, statistics and can be easily controlled remotely
Power consumption	High	Low
Security	Low, voice in analog transmission is transmitted in its native format and can be easily tapped	Very high, encrypted information can be deciphered only by the receiver
Distance covered	Analog signals fade very quickly and hence require repeaters/amplifiers	Digital signals cover large distance and hence require fewer repeaters/amplifiers
Signal reconstruction	Analog signals are difficult to reconstruct as the signal varies and can take any shape	Digital signals can be easily reconstructed as they contain either high (1) or low (0)
Unit of measurement	Hertz (Hz)	Bits per second (bps)

Table 3.2 – Comparison of analog and digital transmission

3.9. Modulation

In telecommunications, every signal transmitted through a medium is modulated. Modulation is the process of manipulating the frequency or amplitude (signal strength) of an incoming signal (voice, video or data) using another high frequency signal called a carrier. This simple technique converts regular signals like low frequency voice into a frequency band that is right for the medium. For example, in cell phone communication, the voice signal is converted into 800–1900 MHz (radio band) for transmission over the airwaves.

3.9.1. Why Modulate?

Modulation is required so that:

- Signals of the same frequency from different sources don't get mixed up. For example, the frequency of the human voice is in the range of 20 Hz to 20 KHz. Thus, without modulation, voice signals from multiple sources transmitted over the same medium will get mixed up and become unintelligible
- Signals travel further than they would have in their native format. Without modulation, the human voice transmitted over a wireless channel will hardly cover any distance. However, after modulation into a band of 800–1900 Mhz, it can travel for miles.

3.9.2. Modulation Techniques

There are two main techniques for modulation:

- Amplitude modulation – The incoming signal is mixed with the carrier in such a way as to cause the amplitude of the carrier to vary at the frequency of the incoming signal. If there are multiple sources, then each source gets a carrier with a different frequency. The frequency of the carrier remains constant.
- Frequency modulation – In this technique, the incoming signal is mixed with the carrier in such a way as to cause the frequency of the carrier to vary above and below its normal frequency. The amplitude of the carrier remains constant.

3.10. Multiplexing

The capacity of the medium carrying the signal is usually much larger than the signal being transmitted. For example, the capacity of a simple copper wire pair is 1 MHz, but the frequency of the human voice is in the range of 250–3500 Hz only, so in effect only 3250 Hz out of that 1 MHz is utilized. To overcome this underutilization of capacity, a technique called multiplexing was designed. Multiplexing allows a carrier to transmit multiple streams over a line.

There are two types of multiplexing used for electrical signals:

- Frequency division multiplexing (FDM)
- Time division multiplexing (TDM)

3.10.1. Frequency Division Multiplexing (FDM)

With FDM, the medium is divided into channels and each channel is assigned a different frequency (Fig. 3.10). The human voice in each channel is still between 250 Hz and 3500 Hz, but the signal in each channel is modulated to a different frequency called a carrier frequency. In simple terms, modulation involves converting the signal from one frequency range (usually lower) to another (usually higher). For example, the voice signal (250–3500 Hz) in the first channel can be modulated using a 10000 Hz signal to convert it to 10250–13500 Hz. The signal in the second channel can be modulated using a 20000 Hz signal, converting it to 20250–23500 Hz. This technique allows a simple pair of copper wire that has a capacity of 1 MHz to carry more than one channel, with each channel receiving its own frequency range.

FDM is only suitable for multiplexing analog signals, and was used extensively in the past. Today, most transmitted signals are digital, so another technique called time division multiplexing (TDM) is used.

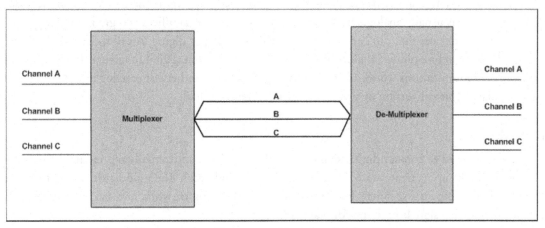

Figure 3.10 - Frequency Division Multiplexing

3.10.2. Time Division Multiplexing (TDM)

With TDM, each channel is given a short duration of time during which it can transmit its signals (Fig. 3.11). Multiple channels are connected to a TDM multiplexer, and each channel is given the entire bandwidth for a short duration. At the receiving end, another device called a de-multiplexer is in step with the transmitter so that the signals from channel A from the transmitting end are sent to channel A at the receiving end. This is accomplished by using a synchronization channel between the transmitting multiplexer and the receiving de-multiplexer. A fixed pattern is sent to the receiver for it to be in step with the transmitter.

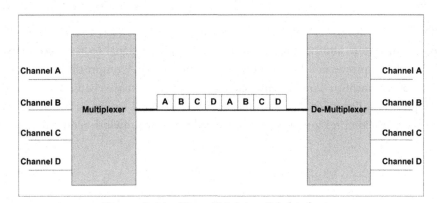

Figure 3.11 - Time Division Multiplexing

The drawback of TDM is that if a transmitter has nothing to send then that time slot is wasted. To overcome this, a new technique called statistical time division multiplexing allocates unutilized channels to transmitters that need it.

3.11. Signal Conversion

Ever wondered why you need a modem to connect your computer to the Internet, whereas phones can be hooked up directly to the phone jack? As you know, the copper wire connection between homes and the central office (local loop) is analog, but computers transmit and receive data in digital format. Hence, they need a device that can convert the digital signals to analog while transmitting, and vice versa while receiving. In general, regardless of whether the application is voice or data, an end-to-end communication channel is established using both analog and digital networks. The local loop is usually analog, and the network between central offices is digital. Similarly, a consumer has devices that can communicate in either analog or digital. A call from a VoIP phone (digital) to a wireline phone (analog) also needs conversion from digital to analog. In today's complex telecommunications world, there is lot of signal conversion between analog and digital formats taking place. This conversion of signals is facilitated by a modem.

3.11.1. Modem

A modem is used to convert digital signals to analog and back and actually represents the two devices that make up its name – a MOdulator and a DEModulator. The modulator is used to convert digital signals into analog signals for transmission on an analog network, and a demodulator is used to convert analog signals back into digital format.

3.11.2. Optical Converters

Another type of signal conversion is from electrical to optical. Most trunk lines, especially between large cities, are optical, and hence all the electrical signals (both analog and digital) have to be converted into optical signals. Electrical signals are converted into optical signals (light or no light).

Signal conversion techniques and equipment have been a great boon for telecommunications, fuelling the growth of new devices (digital) without requiring carriers to change the base network (largely analog). The trend amongst all of the world's leading carriers is to move towards an all-digital all-optical network.

3.11.3. Codec

In addition to converters, another important device that makes signal transmission possible is a codec. A codec COmpresses and DECompresses a digital signal. Compressing a digital signal allows more signals to be transmitted on the same channel. They are extensively used for high bandwidth requiring applications like video. For example, a raw video signal requires a bandwidth of at least 200 Mbps, but with compression, it is possible to reduce the bandwidth required to around 2 Mbps.

4. Basic Concepts of Wireline Voice Communication

The wireline carriers provide voice, video, and data services via wires that connect a customer's premises to a central office. The wireline carriers were the first telecommunications service providers, and have been in business ever since the invention of telephone. For years, they primarily provided voice services, but now also provide video (IPTV) and data (dial-up, broadband) services.

Wireline service providers like ATT, BT, Verizon, Telstra, Bell Canada, and NTT are some of the biggest names in the telecommunication business. For decades, the flagship service provided by these companies was voice through the wired telephone commonly found in many households. However, in most developed countries, people are steadily switching off their wired telephone and moving towards alternatives like VoIP (voice over IP) and mobile. This is seriously denting the main business of these providers. However, these big companies are taking the change in market dynamics seriously. Each of these companies has ventured into providing other services like broadband (DSL) and video (IPTV). In addition, almost all have either a mobile arm or a strong association with a leading mobile service provider.

The focus of this chapter is on wireline voice services only. Chapter 5 deals with wireline data services and chapter 9 deals with wireline video (IPTV) services.

4.1. How do Wireline Phones connect?

We have grown very accustomed to the availability of many things in life; a working telephone is one of them. We pick up a telephone, listen to the dial tone, and dial the number of the person we want to contact without ever wondering how the two telephones, which may be thousands of miles apart, are actually connected. Granted, a simple pair of wires connecting the two telephones would suffice for them to communicate, but with millions of telephones around the world, connecting all of them with each other is impossible. Each telephone would end up having millions of incoming lines.

The overcome this problem, each telephone is connected to a telephone exchange or central office. There is usually one or more central office per city with each having capacity in the order of 10000 lines. All the central offices around the world are in-turn connected through a hierarchical mechanism (refer section 4.2) and thus it is possible to reach any phone in the world.

Each telephone is connected to the central office using a pair of copper wire. This pair of copper wire connecting the telephone to the central office is called the local loop. However, to make things more orderly, the pair of copper wire is not always directly connected to the central office. Large numbers (50 or more) of such pairs are first aggregated within a box called a cross connect box that you see in your neighborhood. Each cross connect box has the capacity to take in 50 or more input copper wire pairs. From the cross connect box runs a thick cable usually containing 50 or more pairs of copper wire to the main distribution frame in the central office, as shown in Fig. 4.1.

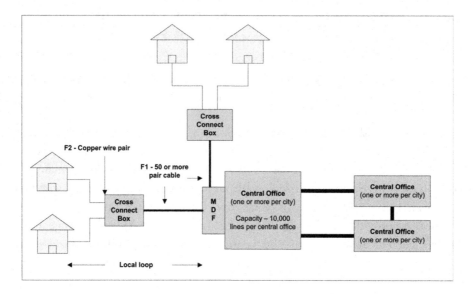

Figure 4.1 - How Do Phones Connect?

Over the last few years, the capacity of the wires connecting customer premises to central office has been increasing. Copper wire is now being replaced by a strand of optical fiber, which creates the huge bandwidth necessary for a high-speed video and data service. Various projects like FTTP (fiber to the premises) and FTTN (fiber to the node) have been undertaken in order to get high capacity fiber into or closer to the premises.

Alternatively, in some countries, copper wire is eliminated altogether and replaced by wireless connection to a nearby tower that is connected to the central office. This type of configuration is called a wireless local loop (WLL) and is quite commonly found in developing countries like India.

4.2. Types of Central Offices

The central office or telephone exchange is the heart of the telecommunications business. It has equipment capable of receiving the signals from telephones at customer premises and determining where the destination telephone is located. Depending on the destination, the call is routed through the following types of central offices:

4.2.1. Tandem Exchange

If the destination is within the same metropolitan area, the call is routed through a tandem exchange. For example, calls between 314-235-xxxx and 314-476-xxxx, as shown in Fig. 4.2, are routed through Tandem Exchange 1. In the US, the first three digits represent a metropolitan area (e.g. 314 for St.Louis), the next 3 represent the exchange (e.g. 235 for Creve Coeur, St.Louis) and the last 4 are unique number assigned to a customer.

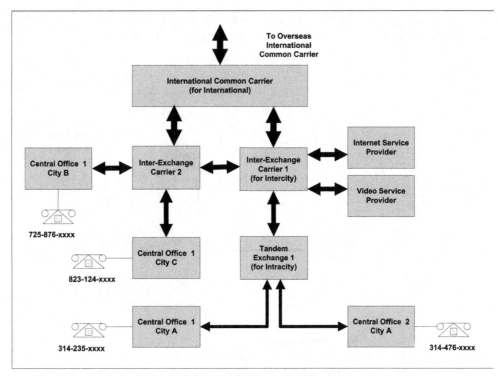

Figure 4.2 - Central Office

4.2.2. Inter-Exchange Carrier (IXC)

If the destination is in a different metropolitan area, the call is routed through an inter-exchange carrier or class 4 exchange. IXCs are responsible for long-distance connections within a country. For example, calls between 725-876-xxxx and 823-124-xxxx are routed through Inter-Exchange Carrier 2 as shown in Fig. 4.2.

4.2.3. International Common Carrier Exchange (ICC)

If the called telephone is in a different country, the call is routed from the class 4 exchange to the international common carrier (ICC) exchange. ICCs are responsible for connection between countries.

4.2.4. Remote Terminals (RTs)

The maximum distance that a local loop can cover is about three miles (five kilometers). This is due to the small resistance that copper offers to the flow of electricity through it, which makes electrical signals fade as the distance increases. Thus, central offices traditionally have a serving area of a three-mile radius around them, but building a central office every three miles is not financially feasible. Each central office usually has a capacity to support 10,000 lines, and it is highly unlikely that there will be that many users within a three-mile radius. It is also not possible to set up a new central office to serve an upcoming sub-division or small village where the number of subscribers is insufficient. Another situation might be that the existing capacity of a central office has already been used up, and it does not make sense to wait until a new central office is opened, or even to open a new one just for a few additional connections. The way to extend the capacity and reach of central offices beyond their limitations is by using remote terminals (RTs) or digital loop carriers. RTs (Fig. 4.3) are low-capacity switches located in small huts or controlled environment vaults. They do not have the full capabilities of a central office, but have the ability to provide a local telephone service on the loops connected to it. The remote and the loops are connected back to the nearest central office using fiber or radio.

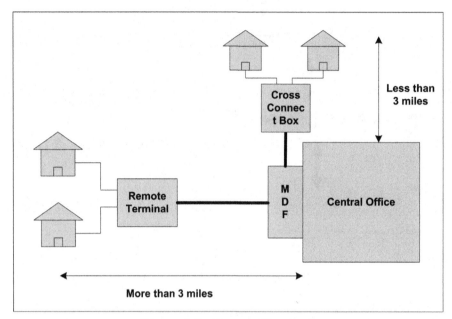

Figure 4.3 - Remote Terminal

4.3. Basic Terminology

4.3.1. Signaling Protocol

Wires and switches only provide connectivity between two telephones. However, to use a metaphor, if two people are in the same room, there has to be a common language for the two parties to communicate effectively. Signaling protocol is that common language, used for establishing, maintaining and tearing down communication sessions between equipments on the telecommunication network. It is classified into two categories: line side, which is between end user equipment and the central office; and trunk side, which is between central offices. Without signaling protocols, the equipments will not be able to communicate with each other and establish a session.

4.3.2. DTMF

DTMF (dual tone multi-frequency) signaling protocol is a line side protocol used for communication between a telephone and the central office. A telephone uses DTMF signaling, which assigns a specific frequency to each key on the pad so that the number dialed can be easily identified by the switch at central office.

4.3.3. SS7

Signaling system 7 (SS7) is the most commonly used trunk side signaling protocol on the PSTN (public service telephone network). The standard defines the procedures and protocol by which network elements exchange information over a network for call setup, routing, and control. Key functionality provided by SS7 are basic call setup, maintain, and tear down (on the trunk side); toll-free (800/888) and toll (900) services; and features such as call forwarding, calling party name/ number display, and three-way calling.

SS7 is also used in linking VoIP traffic to the PSTN network, and in mobile cellular telephony networks like GSM and UMTS for voice and data applications.

4.3.4. PSTN

The network built by the world's wireline carriers is referred to as PSTN (public service telephone network). PSTN is formed by connecting every central office or telephone exchange around the world by coaxial or fiber cables. PSTN is still mostly analog, but many parts of it are being converted to digital.

4.3.5. POTS

POTS (plain old telephone service) represents the standard wireline phone that serves most homes and enterprises around the world. It has been in use for more than 100 years but is being seriously challenged by mobile and VoIP (voice over IP) phones.

4.3.6. IN

Intelligent network (IN) is the network architecture used today for both fixed and mobile telecommunication networks. Before IN, service providers were restricted in their ability to provide value added services (VAS) (e.g., voice mail), requiring the introduction of new equipment to the core switches of the network, which was quite risky. Even adding simple services like call waiting or call forward required a lot of investment, time, and risk. In addition, all the additional equipment had to be purchased from the same switch vendor, which was not a viable business model as the provider was locked into using the same vendor for all enhancements. To overcome these challenges, the concept of intelligent network (IN) was conceived. It envisioned a service-independent telecommunications network, whose aim was to take intelligence out of the switch and move them into modular nodes that could be plugged in without making any significant change to the core switches. In addition, IN forced vendors to build their wares to industry standards, which meant that equipment from different vendors could easily interoperate. This provided the service provider with the means to buy or develop value added services that could be rapidly introduced into the existing network, and easily customized to meet individual customer's needs without using the vendor's help.

The concept of IN sounds very good, but unfortunately it failed to live up to the expectation of enabling hundreds of value added services. The single biggest reason for its failure was that even though the interfaces on the equipment were open, it was still very complicated to carry out any meaningful development. Nevertheless, IN is a significant development in the world of telecommunications that has set the stage for some of the upcoming standards like IMS (IP multimedia sub-system).

4.4. Basic Components

4.4.1. Telephone

A simple telephone consists of a transmitter, receiver, duplex coil, hook switch, ringer, and dialer. The transmitter and the receiver are located in the handset, while the rest is located on the base. Modern telephones have become very sophisticated these days and have features like LCD display, voice mailbox, and so on, but a simple telephone constructed in the early 1900s would still work today.

The transmitter is similar to a microphone, converting the human voice into electrical signals for transmission over the wire. The receiver is similar to the speaker, and its job is to convert electrical signals into audio. The hook switch was originally meant to connect and disconnect the telephone from the network, but these days it is replaced by buttons. The duplex coil is to block you from hearing your own voice, but replaced these days by advanced noise canceling integrated circuits (IC).

4.4.2. Switch

Refer section 3.4 for details on switches.

4.4.3. Transmission Media

Refer section 3.7 for details on transmission media.

4.5. Voice Services

Wired voice services can be classified into two categories based on the customer base they serve:

- Residential
- Enterprise (includes small/medium/large businesses, charitable organizations, and government)

4.5.1. Residential

Wired voice services used by residential customers are few and very simple, consisting of a base wired phone with few additional features. The following are some of the services provided by leading service providers:

- Local
- Long distance (within a country)
- International

The value added services (VAS) for residential customers are as follows:

- Caller ID
- Call waiting
- Call waiting ID
- Privacy manager
- Three-way calling
- Call return
- Call forwarding
- Call screen
- Priority ringing
- Repeat dialing
- Speed calling

4.5.2. Enterprise

In contrast to the simplicity of voice service to residential customers, the voice services to enterprises are quite complex. The following are some of the services:

- Local
- Long distance
- International
- Toll-free (800, 888, 877)
- Direct toll-free service
- International toll-free
- Corporate calling card
- Conferencing
- Messaging
- PBX
- Centrex

The value added services for enterprise customers are as follows:

- Automatic call distributor
- Automated directory services
- Automatic ring back
- Call accounting
- Call forwarding
- Call hold
- Call pick-up
- Call transfer
- Call waiting
- Conference call
- Custom greetings
- Speed dialing
- Busy override
- Direct inward dialing
- Do not disturb (DND)
- Follow-me
- Interactive voice response
- Music on hold
- Welcome message

4.6. Enterprise Voice

Unlike residential customers who mainly require one or two phone lines, enterprises require hundreds or even thousands of connections. Instead of taking thousands of lines from a service provider, enterprises find it cheaper and better to have their own telephone exchanges. As such, there are two types:

- PBX
- Centrex

4.6.1. PBX

A private branch exchange (PBX) is a telephone exchange set up, privately owned, and operated by the enterprise itself. The biggest benefit of owning a PBX is that it is much less expensive than connecting an external telephone line to every desk in the company. However, it is important to note that enterprises owning a PBX have to maintain and operate the branch telephone exchange with their own resources. This is difficult for smaller companies as they do not have the necessary skills or resources. In order to overcome these difficulties, service providers have started offering a hosted PBX, whereby the equipment is owned and maintained by a service provider but still resides in the customer's premises.

4.6.2. Centrex

With Centrex, on the other hand, the service provider owns and maintains all the necessary equipment within their premises. This is suitable for smaller companies who need fewer lines and don't have the resources to manage a complex communication system on their own.

4.6.3. Centrex or PBX - Which Is Better?

The debate over whether Centrex or PBX is a better solution is age old and there is no clear winner. In general, both solutions have their advantages and disadvantages; therefore, every enterprise has to evaluate both solutions against their requirements and capabilities before deciding. Here are some of the advantages and disadvantages of each solution:
PBX advantages:

- Rich features
- Users have administrative control. Changes like adding or dropping a line cost nothing
- System upgraded at user's will
- One-time costs (can receive depreciation benefit), and the equipment belongs to the enterprise thereafter

PBX disadvantages:

- Difficulty managing unrelated business activity
- Need additional employees to manage the systems
- Businesses are not prepared for any communication-related disaster. They often wait until problems occur to carry out repairs
- Upgrades are expensive

In general, PBX is more suitable for large companies who have the ability to manage and maintain a complex telecommunications infrastructure.

Centrex advantages:

- Low initial investment as most equipment required is at the provider's central office. Users pay a small monthly fee
- Services work out cheaper as equipment and staff is shared between many users
- Equipment is maintained and upgraded by the service provider
- Maintenance is covered 24 hours a day

- The service provider is far better prepared to handle any disasters with standby emergency equipment
- Allows enterprises to focus on the core activity and not spend time and money on unrelated activities

Centrex disadvantages:

- Limited feature availability
- Users have limited flexibility. Regular changes like adding or dropping lines cost hefty fees
- The service is only as good as the service provider
- Users pay fees over their lifetime

Centrex, in general, is more suitable for small and medium-sized companies who do not have the capacity to manage and maintain a complex telecommunications infrastructure.

4.7.   Wireline Voice @ 2010

Even though there seems to be a strong market for wireline voice services in developing countries, the heady days are all but over in most parts of the developed world. In many countries, people are 'pulling the plug' and using either a mobile or a VoIP phone in its place. It is true that neither the wireless phone nor the VoIP phone is as reliable as a wireline phone (remember that it works even if there is no power!), but the features and value added services (VAS) on a wireline are very limited and there is very little chance of improvement. It is very likely that the demise of wireline phone will have started by 2010, and there will be a real fall in numbers from the current 1.85 billion around the world. For example, US RBOCs (Regional Bell Operating Companies – e.g. SBC, Bell South) lost 150,000 subscriber lines per month in 2007. At the same time, VoIP service providers are adding about 100,000 subscribers per month. The balance of local service subscription losses of about 50,000 are moving to wireless-only plans or canceling their secondary household lines.

5. Basic Concepts of Wireline Data Communications

The second of the six big revolutions in telecommunications came about with the advent of technology to exchange data between machines. Voice communication started the era of instant communication and fueled the need to not only talk, but also exchange data. By the late 1960s, the ARPANET team successfully connected two computers and exchanged data between them. Around the same time, companies started investing heavily into using computers for data processing and running other business applications. Computers were coming in different sizes and varying capabilities. The mainframes were the workhorses, and other medium and small computers picked up the rest of the load. With so many computers around, each specializing in specific tasks, connecting and exchanging data between them not only seemed logical, but also became necessary. Thus began the era of data communication.

Companies started installing computers at a single location, but quickly moved to interconnect with those located at other locations and also making them accessible to users through smaller desktop computers. For e.g., the mainframes were usually located in a climate-controlled office at one location, but were needed by users spread across the country. The growth of the Internet added another dimension to data communications; individuals with computers at home also began communicating with other computers for both personal and business communication. The structure thus formed by interconnecting computers and other electronic devices is referred to as a Network.

5.1. What is a Network?

A network is defined as a group of electronic devices like computers, printers, and servers connected with each other by a communication channel formed by cables or free space (wireless). A network allows machines to share and exchange information like files, databases and programs, access common services like email, and share hardware.

5.2. How is a Network formed?

At the minimum, a network consists of the following components:

- Communicating devices – The network contains at least two or more devices that would like to communicate with each other. It can be two computers, a computer and a printer, or any other combination of devices.
- Protocol – Protocol is a set of rules used during communication; it is like the language for communication. Typical examples of protocol are Ethernet, ATM, frame relay, and IP.
- Medium – A medium is the communication channel that connects the communicating devices. The medium can be a simple copper wire, CAT 5 cable, or free space (wireless).
- Router – Two computers can be connected to each other by a simple wire, but when more than two computers are involved, a hub (for small networks) or a router (for large networks) is required. This device acts like a switch and connects two communicating devices, no matter where they are physically located.

5.3. How does a Network work?

Fig. 5.1 shows how two computers communicate with each other. The network shown below represents a typical small office or home network with a few computers and a printer.

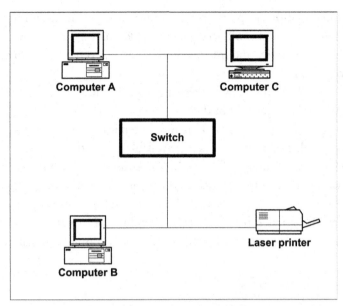

Figure 5.1 - A Simple Network

If computer A wants to talk to computer B, the following steps take place:

- Computer A drafts a message, which contains the address of computer B and the message it wants to convey. This is like sending a letter with the address written on the envelope.
- The switch is like the post office, which receives the message and figures out from the destination address of the message that it is meant for computer B. It also knows how to connect to computer B.
- The switch forwards the message to computer B.
- Computer B sends acknowledgment of receipt of the message from computer A back to the switch.
- The switch relays the acknowledgment to computer A.

However, real-life communication between two devices is far more complicated than this. The systems connecting the two devices take care of security, protocol translation if required, and the message is sent again if delivery fails.

5.4. Advantages of Networks

Networks enable computers to connect and communicate with each other no matter where they are located. Some of the key advantages of having a network are as follows:

- Speed – Networks provide a very rapid method for sharing and transferring files. Without a network, files are shared by copying them to disk, then carrying or sending the disk from one location to another. This method of transferring files is very time-consuming
- Costs savings – Network-based communication, especially emails and instant messages save a lot of money. For enterprises, network-based versions of software are relatively cheaper than individual licenses. In addition, enterprises benefit greatly from having centralized file and database systems rather than having separate systems at each location
- Integrated workforce – Many enterprises are global, therefore it is important to connect offices

and people located in different parts of the world in order to create an integrated workforce. The benefits of an integrated workforce are many, like sharing knowledge, taking advice from experts, making collaborative and informed decisions, less travel, and providing more visibility and control to upper management

- Resource sharing – Networks allow the sharing of critical resources like programs, equipment and data between users, regardless of their location
- Failover resilience – Networks allow companies to replicate critical data at multiple locations. If one location is disabled then the data is still available from other location
- Security – The biggest advantage of closed networks used by enterprises is that they protect not only the hardware and software, but also highly proprietary data contained within the network
- Centralized administration – Enterprises with a large number of computers cannot maintain them (like software upgrades, problem fixes) individually. Networking technology allows routine maintenance from a single point, affecting a large section of the computers at each time

5.5. Challenges of Networks

- Initial costs – Networks save a lot of money over time, but they require a huge upfront investment that many smaller companies may not be able to afford
- Maintenance – Networks are inherently complex and require regular maintenance. The recurring cost involved in maintenance is also quite high
- Security – Large institutional networks unfortunately become visible targets for hackers. Top networks that attract the attention of hackers usually belong to government departments like defense, internal security, and other high profile enterprises in the areas of pharmaceuticals, chemicals, and software
- Failure – Networks are inherently secure, but occasionally you hear a bunch of people having fun in a coffee shop during work hours because their network is down. Networks do not go down often, but when they do, they affect a number of computers and equipment, thereby paralyzing large portions of the workforce

5.6. Basic Components

Many components other than just computers and printers make up a network. This section describes many of those components (Fig 5.2). However, it is important to note that not all components are used in all networks.

5.6.1. Network Interface Card (NIC)

A NIC is used to connect a device to the network. It is designed for a particular type of network (LAN, WLAN, MAN, WAN), protocol (Ethernet, token ring, ATM, frame relay), and media (twisted pair, coaxial, or fiber optic cable). It is a combination of hardware and software.

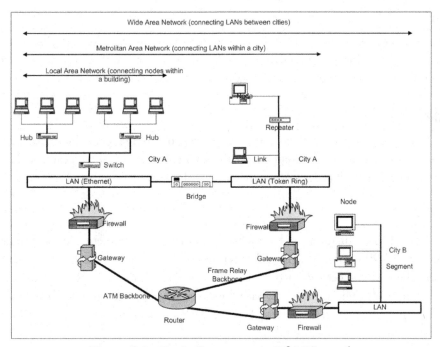

Figure 5.2 - Basic Components Of A Network

5.6.2. Modems

Modem is short for two devices – MOdulator and DEModulator. A typical large network consists of digital and analog equipment. Modem helps bridge the connection between analog and digital networks by converting the signal as required. Typical use of a modem is to connect a computer to the telephone line in order to connect to the Internet while using a dial-up or DSL connection. Computers communicate in digital format, while the telephone line is capable of supporting analog only and hence a modem is required for a computer to use telephone line as communication channel.

5.6.3. Backbone

A backbone is a high-capacity wire that connects multiple parts of a network. For e.g., multiple local area networks (LANs) are connected together by a backbone to form a metro area networks (MAN).

5.6.4. Hub

A hub is a common connection point for devices in a network. Multiple segments are connected to a hub, which is in turn connected to a router, switch, or backbone.

5.6.5. Repeater

Repeaters are used to connect two or more segments. Depending on the media type, the protocol used, and the switching technique, there is a limitation on how far (distance) a receiving device can be placed from the source. For example, the maximum length supported in a LAN running 10BaseT Ethernet is just 100 m. If nodes are located beyond such a limit, repeaters are used. Repeaters restore the signal by amplifying them and filtering out any unwanted noise.

5.6.6. Switch

A switch acts as a connection point for different nodes in a network. This is similar to a hub, but unlike a hub, which broadcasts all messages to all devices, a switch only sends the message to the destination devices(s). In addition, they are faster and smarter than hubs, but more expensive too.

5.6.7. Bridge

A bridge is a device that connects two LANs. They are protocol independent, and hence capable of connecting LANs that run on different protocols. A basic bridge is just a repeater that broadcasts an incoming data stream to all other LANs. They can be thought of as low-level routers.

5.6.8. Routers

In a large network, a destination can be reached by taking many different routes. A router is a device that acts as a decision maker at each intermediate hop on which path to take next. It holds information about all the network paths it is connected to, and uses complex routing algorithms and routing protocols to decide the best path for the incoming packet to take. It knows which paths are congested and which are not working. When a packet arrives, the destination address is used to determine which path is the best to take.

There are three types of routing protocol:

- Static routing protocol is the simplest, and capable of supporting small network
- Routing information protocol (RIP) is capable of supporting medium-sized network
- Open shortest path first (OSPF) is capable of supporting large networks

A router provides far more functionalities than a bridge, but is usually slower and more expensive.

5.6.9. Gateways

A gateway is a hardware device or software program that is used to translate between incompatible protocols. They can translate packets coming from Ethernet based LAN to ATM packets ATM based MAN or WAN.

5.6.10. Firewall

A firewall is used to secure networks. It is located at the gateway, and protects access to resources (information and devices) located within a network from other networks. All incoming and outgoing data is checked and dropped if it is malicious or unsolicited. They are difficult to configure properly, but are the best defense available today.

There are two types of firewalls:

- Hardware firewall – A network or hardware firewall is usually located at the edge of the local network so that all traffic (packets) flowing through are properly scrutinized. Hardware-based firewalls are usually used by enterprises that need to protect a large number of computers and resources
- Software firewall – Software-based firewalls are run on devices that need to be protected.
In general, hardware firewalls are more difficult to break into than software firewalls.

5.6.11. Cable

A network is formed by connecting devices with cables (unless it is a wireless network), and requires cables capable of transmitting at different speeds. Cable forming the backbone requires very high bandwidth, while the cable forming the segment requires lower bandwidth. The American National Standards Institute (ANSI) has classified the different types of cables into categories (CAT). Each category has standard features like the speed and the operational frequency range defined:

- CAT 1 – The maximum data rate supported by CAT 1 is only up to 1 Mbps. It is mostly used for voice communication and ISDN service.
- CAT 2 – The maximum data rate is up to 4 Mbps. It is used in token ring-based LANs.
- CAT 3 – The maximum data rate is up to 16 Mbps. It is used in Ethernet-based LANs.
- CAT 4 – The maximum data rate is up to 20 Mbps. It is not used much.
- CAT 5 – The maximum data rate is up to 1000 Mbps. It is used quite commonly at home and enterprises as cable to connect devices.
- CAT 5e – The maximum data rate is up to 10 Gbps. It is used in advanced versions of Ethernet (10 Base-T or 100 Base-T) and ATM.

All cables have certain limitations, and they have to be fully evaluated against the requirements before use. Following are some of the factors to be kept in mind during evaluation:

- Attenuation – This is the measure of decrease in signal strength as it travels through a medium. It is usually expressed as the ratio of received signal strength to transmitted signal strength, and is measured in decibels per kilometer (dB/km) at a specific wavelength or frequency. The lower the number, the better the cable is. Attenuation is heavily dependent upon the frequency of the transmission - the higher the frequency, the higher the attenuation. It is also dependent upon the physical characteristics of the material used to make up the medium.
- Cross talk – Cross talk occurs when signals running on adjacent wires interfere with each other. This unwanted interference distorts the signals.

5.7. Basic Terminology

Simple networks connecting few computers within a small building are not complicated, but understanding large networks spanning the globe is quite difficult. Some of the terminology used in networks is explained in this section (Fig 5.3).

5.7.1. Node/Device

Any communicating device connected to the network is called a node or device. Computers, servers, switches and hubs are good examples of nodes.

5.7.2. Segment

A segment is a section of a network that is physically separated from the rest of it by a hub, switch, bridge or router.

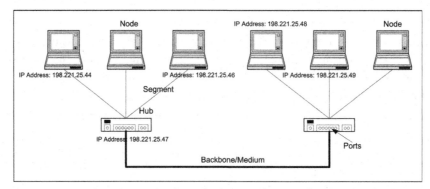

Figure 5.3 – Network Terminology Reference

5.7.3. IP Address

An IP address is a unique identifying number, much like the postal address of a house, that every node on the network will have. It is used by all devices to identify and communicate with each other. A typical IP address following IPv4 standard is 32 bit, and in human readable form looks like 316.443.678.2. In all, there are about 4.3 billion IP addresses in IPv4, but the world is fast running out of them. The next version of the IP addressing standard called IPv6 is in place, but is still not very widely used. The main difference between IPv4 and IPv6 is that IPv6 uses 128 bits to address each node. This provides an extremely large number of unique addresses (340,282,366,920,938,463,463,374,607,431,768,211,456).

IP addresses are managed and created by the Internet Assigned Numbers Authority (IANA). IANA generally assigns super-blocks to regional Internet registries, who in turn allocate smaller blocks to Internet service providers and enterprises.

5.7.4. Ports

There are two types of ports on a computer:

- Hardware port – This is an interface on a node/device to which wire coming from other node is plugged in
- Software port – This is a virtual data connection that can be used by applications to exchange data directly. Each application is assigned a port so that traffic from that application is routed through that virtual port.

5.7.5. Frame

A frame is the basic unit of data sent from one port to another. Depending on the protocol, a frame can be hundreds or even thousands of bytes long. It is composed of a starting delimiter (SOF), a header, the payload (data), error-checking codes (CRC), and an ending delimiter (EOF). It is used as the basic unit of data in layer 2 and layer 3 (LAN, WAN) protocol, while a packet is the term used for the basic unit of data in higher layered protocol like TCP/IP.

5.7.6. Cell

A cell is a fixed-size packet that contains control and data information. It is the basic unit of data in ATM protocol and is 53 bytes long. In comparison, frames and packets are of variable lengths.

5.7.7. Packet

A packet is a unit of data sent over a network, as measured in higher layered protocol like TCP/IP. Each packet contains information about the sender and the receiver, error-control information, and the actual message. Packets may be of fixed or variable length.

5.7.8. Network Topology

Network topology is a general term for the different ways (Fig 5.4) that nodes are physically connected in the network. There are five basic types of network topology:

- Bus: in this type, all the devices on the network are connected to a common cable. However, the ends of cable are not connected together, but end on a terminating device
- Ring: in this type, all the devices on the network are connected to a common cable. In addition, the ends of cable are connected together to form a loop
- Star: in this type, the topology has a central connector to which all the devices are connected
- Tree: in this type, nodes are connected in hierarchy with a root node at the top
- Mesh: in this type, each device is physically connected to every other device. This topology is good for small networks, but not practical for large ones.

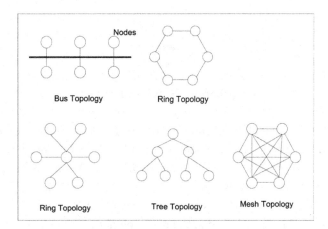

Figure 5.4 - Network Topology

5.7.9. Media Access Control (MAC)

Media access control (MAC) is the hardware address that uniquely identifies each node in a network. It is six bytes long, with the first three bytes identifying the manufacturer, and the remaining three bytes a unique number assigned to each device by the manufacturer.

5.8. Network Layers (OSI Model)

Data communication activities like sending an email or browsing the Internet involve very complex interactions amongst various devices and the network. The complexity comes from the fact that each computer may use different applications to generate or read the data, they may be part of networks running on different protocols, and the data might pass through copper wire, wireless, or optical fiber. The combination created by these varying factors is very complex. To understand and simplify the complexity, several theoretical models representing a network have been developed. The approach of each of these models is to divide the complex process into smaller sub-processes and standardize the activities performed by them.

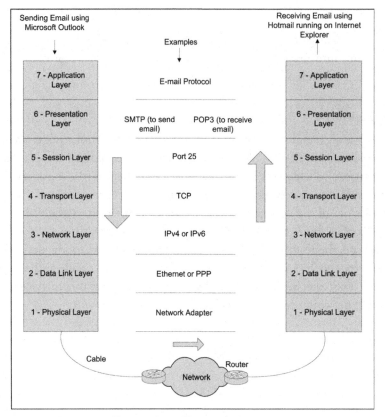

Figure 5.5 - Network Layers

The OSI (open system interconnection) model is the most popular networking model from a theoretical perspective. It acts as a reference model showing how messages must be transmitted from one computer or node to another. The OSI model divides the communication process into seven distinct layers (Fig. 5.5), with each layer having clearly demarcated roles and responsibilities. Every message sent from one node to another goes through each of these layers. At the sending node, the message goes from layer 7 to layer 1, and at the receiving node, the message goes from layer 1 to layer 7. The implementation of each of these layers is carried out by a combination of hardware and software running on each node. The seven layers in detail are as follows:

5.8.1.   Application Layer - 7

The primary job of the application layer (referred to as layer 7) is to support application programs like Internet Explorer or Outlook (for email). It is not the application itself but is a service provided to the application that is directly used by a user. It authenticates the sender (or receiver), makes sure that the resources are available to establish the session, and conducts communication.

5.8.2.   Presentation Layer – 6

The primary job of the presentation layer (referred to as layer 6) is to convert the data received from the application layer into a format that the network expects. On the receiving end, the presentation layer converts the data received into a format that the application layer expects. In addition to conversion, this layer also provides encryption so that the data sent over the network are secure.

5.8.3. Session Layer - 5

The primary function of the session layer (referred to as layer 5) is setting up, maintaining, and gracefully terminating the session at the end between the two communicating nodes.

5.8.4. Transport Layer - 4

The primary function of the transport layer (referred to as layer 4) is to ensure the completeness of the received data. It determines whether all the packets have arrived, detects errors, recovers lost data, and manages the retransmission of data. Transmission control protocol (TCP) is a good example of layer 4 protocol.

5.8.5. Network Layer - 3

The primary job of the network layer (referred to as layer 3) is switching and routing. It creates and establishes logical paths (virtual circuits) between two communication nodes so that data can be transmitted between them. It determines the address of the next node in the communication path (remember data can pass through multiple nodes like routers and switches before reaching the destination). This layer is also responsible for error handling, path congestion management, and packet sequencing. A good example of the network layer is Internet protocol (IPv4 and IPv6).

5.8.6. Data Link Layer - 2

This primary job of this layer (referred to as layer 2) is converting data between frames and bits. On the sending side, it receives data in frames and converts them into bits. On the receiving side, it converts bits coming in from the physical layer into frames. In addition, it handles errors in the physical layer, flow control, and frame synchronization. Ethernet is a very good example of the data link layer.

5.8.7. Physical Layer - 1

This primary job of this layer (referred to as layer 1) is to establish and teardown the circuit established between two adjacent communicating nodes. In addition, this layer also makes provisions for monitoring and reporting certain parameters, such as bit error rates or failures via a management system. DSL is a good example of the physical layer.

5.9. Network Classification

Networks have become enormously complicated over the years. To make sense of the complexity, networks are usually classified based on either the size or the protocol used to build them. Some other forms of classification are based on the type of hardware used, the access and security type, and so on.

In terms of size, a network can be classified as follows:
- Personal area network (PAN)
- Home area network (HAN)
- Local area network (LAN)
- Metropolitan area network (MAN)
- Wide area network (WAN)

Networks classified based on the protocol they use are discussed in next section (5.10).

5.9.1. Personal Area Network (PAN)

A personal area network (PAN) is the network formed by connecting devices within the range of an individual person, usually within 10 meters. Typical devices that are part of a PAN are laptops, PDAs, cell phones, or other handheld devices. The advantage of PAN is that no wires are required to interconnect these devices in order to allow them to share and exchange information. PAN, or wireless PAN, is usually enabled by a popular technology called Bluetooth. Bluetooth was first crafted by Ericsson, but later adopted by a host of other manufacturers like Nokia, Toshiba, Intel and IBM as the standard for PAN.

Bluetooth uses the radio waves located in the frequency band of 2.4 GHz and transmits voice and data at less than 1 Mbps. A PAN formed using Bluetooth is called piconet and has the capacity to allow the interconnection of up to eight devices at any given time. Devices in one piconet can talk to devices in other piconets or connect to nearby hotspots. The most common application of Bluetooth is connecting cell phones to wireless headsets, but it is increasingly finding use in many other applications as well.

5.9.2. Home Area Networks (HAN)

HANs are larger than PANs, and as the name suggests are found in homes. Multiple computers within a home connect and share common resources like a printer or broadband connection. They are usually formed by wireless routers (802.11a, 802.11b and 802.11g), Bluetooth or some form of Ethernet protocol (10Base-T, 100 Base-T). They now go beyond connecting just computers and printers, and include entertainment devices like TV, gaming consoles, and video cameras.

5.9.3. Local Area Networks (LAN)

LAN is a non-public network connecting computers that are located within a small geographic area, usually within the same building or campus (see Fig. 5.6). The connected computers share applications, data, data storage facilities, and peripheral devices. Devices forming the LAN can be connected by twisted-pair wire, four-wire pair, coaxial cables, fiber-optic cables, or wireless protocols. LANs support speeds of up to 100 Mbps, and are built to serve a user group whose entire population is related to a single organization (e.g., enterprise, school).

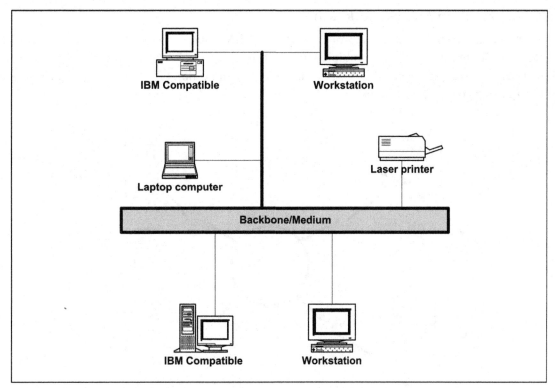

Figure 5.6 - Local Area Network (LAN)

A LAN can serve anywhere from two to a couple of thousand users. In fact, a "Class A" LAN can theoretically connect up to 16 million devices. LAN implementers divide all the users into logical groups called subnets, which might include users in one building or users belonging to a sales organization. The concept of subnets allows LANs to use a single IP address for the entire subnet, thereby saving IP addresses. In addition, a single IP address is used by the LAN for communicating with the outside world, which helps in creating a single secure gateway to the public (Internet) or other private networks.

LANs are established using various hardware and software technologies. The most common software protocols used in establishing LANs are:

- Ethernet
- Token ring

Ethernet is by far the most popular protocol on LAN. The first Ethernet protocol-based connection was designed and tested by Bob Metcalfe at Xerox, who also developed the standards for connecting various devices. Ethernet protocol is explained in much more detail later in this chapter.

Token ring is the second most widely used LAN technology. Devices are connected in ring topology. A token is passed around the ring, and any device that wants to transmit a message to another device on the ring captures that token and puts a frame containing data on the ring. The frame goes around the ring and the receiver to whom the data is intended copies it. The frame is removed when it returns to the sender, and at that time, the sender releases the token for other devices to capture.

5.9.4. Metropolitan Area Network (MAN)

MAN is a data communication network that connects LANs located within a metropolitan area (Fig. 5.7). Enterprises that have multiple offices within a city install LANs at individual locations and connect all those LANs using MAN. The backbone connection for MAN is usually fiber and has a capacity of up to 100 Mbps.

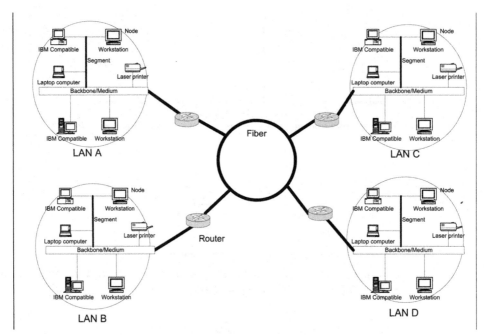

Figure 5.7 - Metro Area Network (MAN)

5.9.5. Wide Area Networks (WAN)

WAN is a data communication network that connects LANs and WANs located in different metropolitan areas (Fig. 5.8). WANs span large distances, and the Internet is in fact the biggest WAN of all, covering the entire world.

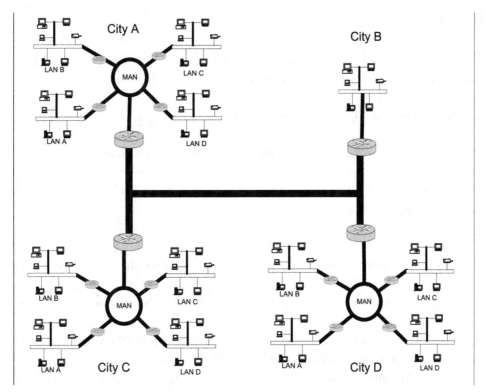

Figure 5.8 - Wide Area Network

The key difference between WAN and LAN or MAN is that WAN is not always owned and operated by a single entity, although there are some instances where big multinational enterprises do have private WANs. In addition, unlike LAN or MAN where the entire network is built using a single protocol, WANs are built with a combination of protocols like frame relay, ATM and MPLS. They are usually owned and operated by telecommunication companies.

5.10. Network Protocols

Protocols are a set of standardized rules for exchanging information among computers, used in a similar way to the language used for communication between humans. Just like any language, they have a precise set of rules and syntax that govern the format of information exchanged. Apart from carrying information, they are used to establish, carry out and terminate a connection. They are also used to coordinate billing and customer care systems, manage network devices, and any other processes that require coordinated communication and control. In general, protocols allow computers and all electronic devices to communicate with each other.

There are thousands of types of protocols, but the following are the most prominent:

- Ethernet
- TCP/IP
- HTTP
- SONET/SDH
- ATM
- Frame relay
- MPLS

5.10.1. Ethernet

Ethernet is the most popular protocol used in local area networks (LANs), with close to 85% of LANs across the world using it. It is not only used in enterprise LANs, but is also quite popular in home networking, and refers to a family of protocols that are defined in IEEE's 802.3.

The popularity of Ethernet is derived from the following advantages:

- Open source standard that is available for free to all device manufacturers who build devices used in Ethernet
- Easy to understand
- Easy to implement and maintain
- Robust

The Ethernet family currently consists of many standards, of which the most popular are:

- 10 Base-T – Provides transmission speeds up to 10 Mbps. It is implemented using twisted pair copper wire or CAT 5
- 100 Base-T (fast Ethernet) – Provides transmission speeds up to 100 Mbps. It is implemented using two pairs of twisted pair copper wire or CAT 5
- 1000 Base-T (gigabit Ethernet) – Provides transmission speeds of up to 1000 Mbps or 1 Gbps. It is implemented either using four pairs of twisted pair copper (CAT 5 only) or optical fiber

5.10.2. TCP/IP

Transmission control protocol/Internet protocol (TCP/IP) is a suite of protocols that is used for communication over the Internet. The TCP part of TCP/IP protocol sends data as an unstructured stream of bytes and monitors the transfer of data. It is capable of requesting and resending lost packets and recognizing duplicates. It also has control mechanisms in place to coordinate communication between a fast and a slow device. In OSI mode, TCP falls in the transport layer (layer 4). The IP part of TCP/IP receives the unstructured stream of data from TCP and converts it into packets. It is capable of providing data security, and reports any errors. In OSI mode, IP falls in the network layer (layer 3).

The advantage of TCP/IP is the flexibility it provides for higher-level protocols like HTTP and FTP to ride on it for communicating with devices using the Internet.

5.10.3. HTTP

Hypertext transfer protocol (HTTP) is a communications protocol used extensively to transfer or convey information on the World Wide Web. The development of HTTP was coordinated by the W3C (World Wide Web Consortium) and the IETF (Internet Engineering Task Force).

HTTP is a request/response protocol between a client and a server. Clients like Web browsers make a request to a server that stores or creates resources such as HTML files and images. HTTP is an application layer (layer 7) protocol that commonly uses TCP/IP.

5.10.4. SONET/SDH

Synchronous optical network (SONET) protocol is the North American protocol for long-haul fiber-optic networks. The data rates supported by SONET range from 51.8 Mbps to 10 Gbps or higher, and the standards for data rates are defined in a hierarchy called optical carrier (OC) levels. The international standard equivalent to SONET is known as synchronous digital hierarchy (SDH), with minor differences between them. These two standards are extensively used by telephone carriers, and it is the protocol of choice for ATM, frame relay, and MPLS to ride over it.

The advantage is the flexibility it provides to ride any higher-level protocols, including ATM and frame relay. The disadvantage is that this protocol is mostly suitable for circuit-switched voice traffic. To support today's IP traffic, packet over SONET and Ethernet over SONET, which make better use of the tremendous bandwidth of SONET-based networks are popular.

In the OSI model, SONET/SDH falls into the physical layer (layer 1).

5.10.5. ATM

Asynchronous transfer mode (ATM) protocol is an international standard for voice and data communications. ATM supports a data communication rate of 155 Mbps to 622 Mbps, with new standards pushing the speeds up to 10 Gbps.

ATM organizes digital data into cells of 53 bytes each. There are multiple advantages of having small and constant-sized cells. A constant-sized cell can be processed by hardware instead of software, which reduces processing time. The small, constant cell size also allows ATM equipment to treat voice and data the same. Smaller packets are particularly suitable for real-time video and audio communication, and the size ensures that the network is not blocked by long packets.

ATM technology is designed to improve the utilization and quality of service (QoS) on high-traffic networks. Without routing and with fixed-size cells, networks can manage bandwidth much more easily under ATM than under Ethernet. However, the high cost of ATM relative to Ethernet is one factor that has limited its adoption mainly to the backbone.

5.10.6. Frame Relay

Frame relay is a high-speed packet-switching data communication protocol used primarily for interconnecting LANs. It transmits data in units called frames, which vary in size but are usually 4,096 bytes. It is quite efficient as it leaves all error-correcting operations to sending or receiving nodes, thus focusing upon transmission and making it very efficient.

Frame relay transmits data by establishing virtual circuits. To an end user, a virtual circuit looks like a dedicated path, but in reality, it is not. A complete path between the two nodes is identified and established before transmission begins. All the packets transferred follow the same route and arrive in the same sequence as sent, giving the impression that a dedicated route is established just for the conversation. The advantage of having a virtual circuit is limited routing is required making it faster and reliable.

The advantage of frame relay is that it is can handle packets of variable length and support other higher-level protocols to ride on it as well. The biggest disadvantage is that it can only support data communication. It is not very well suited to audio and video communications.

5.10.7. MPLS

Traditionally, networks around the world are built and used specifically for a particular type of traffic. For example, frame relay is mostly suitable for data networks while ATM for voice and data only. However, in today's world of IP technology, traffic is made up of different types of signals (voice, video and data). Even though each of these signals is converted into packets, they each need to be treated differently. For example, video is real-time traffic and requires a certain amount of bandwidth to be reserved for an acceptable quality of viewing experience (assured quality of service). Frame relay and ATM are not very flexible to support the bandwidth management and service requirements of this new type of IP-based traffic. MPLS provides a scalable and effective solution to these issues and provides effective traffic management and routing capabilities. In addition, it is capable of easily running on top of ATM and frame relay, thereby making the best use of them.

Multi-protocol label switching (MPLS) was originally presented as a way of improving the forwarding speed of routers, but is now emerging as a crucial standard technology that offers new capabilities for large-scale IP networks. Traffic engineering – the ability of network operators to dictate the path that traffic takes through their network – and virtual private network support are two key applications where MPLS is superior to any currently available IP technology.

5.11. Device Configuration

One of the interesting aspects of Network study involves how devices that want to communicate with each other are set up. There are several types of configurations and each of them explained here behaves very differently.

5.11.1. Client-Server

In a client-server network configuration, a node (client) requests information from another node (server). A simple example would be accessing your bank account online. Your machine (client) requests a large powerful computer (server) sitting in the bank's data center to fetch you the current balance in your account. A single server serves a very large number of clients, and usually uses complex software (code, database) and hardware infrastructure (application servers, Web servers) to perform its duties.

The Internet is a very good example of a client-server model. Almost all the Web sites that provide news, information and e-commerce follow the client-server model.

5.11.2. Peer to Peer

Peer to peer (P2P) is a network architecture where each node (computer) shares a part of its own resources (hardware and software) with other nodes in a mutually beneficial way. In this configuration, every node is both a client and a server. At any point, a node may act as a client by connecting with another host to download information, and a moment later, it may be a server, dispensing information (probably the same) to another host.

Napster and Kaaza are very good examples of P2P, widely used to share music. It is reckoned that at least 60% of the Internet traffic today is P2P, with consumers mainly using it for downloading music, films and software.

5.11.3. Grid Computing

It is a widely known fact that no computer is ever fully utilized. For example, most PC users use only 5% of the machine's capacity at any given time, and most enterprises use less than 40% of their server capacity. There is an enormous amount of unutilized computing capacity, and a grid is a way to overcome this. Using a grid means sharing computer power and data storage capacity over a network. For example, a company's office in San Francisco can utilize spare server capacity when required from its office in New York. In fact, a grid goes well beyond simple communication between computers; its ultimate aim is to turn the global network of computers into one vast computational resource.

5.12. Network Security

Network security is about preventing and detecting unauthorized use of digital assets and computing power. Depending on the type of user (residential or enterprise), the security concerns are different.

Residential customers mainly connect to the Internet using a DSL or cable modem. The biggest concern among residential broadband users is securing their machines from viruses. A great deal of time and money is spent addressing this concern, but there is still no silver bullet. Hackers write malicious programs – viruses, trojan horses, worms, ad wares, spy wares that cruise around the broadband super-highway looking for vulnerable systems. Once a vulnerable system is found, the hacker uses the computing resources of the system to launch spam, store files illegally, and steal critical information like social security numbers, bank account numbers, and passwords. Research shows that broadband users are five times more likely to be targeted by hackers than dial-up users.

The two greatest advantages of broadband – always-on connection and high speed – are the two biggest sources of security problems. The always-on connection feature means that the computer is always accessible for attacks. The high-speed nature of broadband means that hackers can upload critical information and store malicious content on the machines much faster than if the machine is on dial-up. It has been observed that systems with a broadband connection are "sniffed" once every few seconds by hackers for vulnerability.

Fully securing systems while on broadband is not always possible. However, systems can be largely secured by following some simple steps like installing patches and upgrades on the software, and installing a software/hardware firewall. However, no solution is foolproof and new viruses find loopholes that existing protection mechanisms cannot take care of. No solution can provide blanket protection against all viruses all the time. The only way to achieve maximum protection is to regularly update the mechanism used for protection. Certain anti-virus providers publish as many as three or four updates every week, and unless systems are updated, there is no protection. Another aspect that has compounded the security problem is the increasing popularity of broadband among the non-tech savvy population. No amount of additional components like firewalls and anti-virus software are useful if they are not configured properly. The unfortunate part is that configuring a security mechanism is difficult even for the tech-savvy population.

The security requirements of enterprise are far more complicated than the requirements of residential customers using broadband. Enterprises have a lot more data and assets that need to be protected, and a lot more access points from where hackers can intrude into the system. The access points can be a Web page, or hundreds of servers each connected to the network, or the machines that employees use for access while working from home or traveling.

Regardless of the security threat, broadband is here to stay. According to the most recent global statistics from Nielsen NetRatings, the number of broadband connections worldwide has grown 130% in just over a year, and promises to grow even faster over the next decade.

5.12.1. Security Tools

A variety of tools are used for managing security, ranging from hardware solutions like a network firewall to software solutions like VPN, SSL, anti-virus, personal firewall, anti-spy wares, and many more. Regardless of which tool is employed, it is important to note that no single solution provides complete security. Most installations use a combination of the following tools to achieve maximum security:

- Firewall – The most commonly used security tool. The firewalls employed by enterprises are far more powerful than those used by residential customers.
- Virtual private network (VPN) – VPN is the most commonly used tool when enterprises need to allow entities to connect to their network through the Internet. Typical examples are employees working from home, or enterprises exchanging data with partners. Without any security mechanism in place, it is very easy for hackers to manipulate the data exchanged, or even use the connection to get inside the network. VPN allows entities to communicate by establishing a secure and highly encrypted connection between the computers that is very difficult to intrude.
- Secure sockets layer (SSL) – SSL is another widespread mechanism for secure communication, but unlike VPN, which is generally used to extend private networks, SSL is used for secure communication over the Internet. Many Web sites use it when exchanging confidential information like credit card numbers, etc. It creates a secure connection between the computers and any amount of data transferred over it is safe.
- Digital certificates – People only engage in confidential communication with others if they recognize them. Digital certificates work on the same principle - each entity has a digital signature issued by a trusted third party that uniquely identifies them. Before communicating, machines exchange the digital certificate, which ensures the identities of those they are communicating with. This is a very common mechanism used in networks today.

5.13. Residential Broadband

Broadband is a set of technologies that allows the transfer of information at a very high speed. To qualify as broadband, a system must be capable of carrying out transmission at the rate of at least 256 Kbps. These technologies can be wireless or wireline, and include xDSL, cable broadband, 2.5G and higher, and WLANs. Broadband has completely revolutionized the field of telecommunications. Before broadband, telecommunication mainly meant voice communication, but broadband has made a completely new world of communication possible. Using broadband, multiple types of services, like voice, video and data, are all possible on a single medium like twisted pair copper, coaxial cable, or fiber. In addition to these basic services, value-added services like faster access to the Internet, podcasts, music downloads, video on demand, VoIP, Skype, and many more are made possible.

Broadband is one of the fastest growing sectors in the telecommunications market today. The broadband market had a growth rate of around 55% in 2007 and the number of connections is over 500 million worldwide. China, South Korea, Japan, Western Europe, Australia and the USA are all fuelling this tremendous growth.

The most popular residential broadband technologies are as follows:

* Digital subscriber line (DSL)
* Cable broadband
* Wireless 2.5G or higher
* ISDN
* Fiber to the home (FTTH)

5.13.1. xDSL

DSL is the broadband solution offered by telephone service providers. The popularity of DSL comes from the fact that it runs on existing twisted-pair copper wires, which means relatively small investment for tremendous benefits. DSL currently provides download speeds of 1.5 Mbps or higher and upload speeds of 384 Kbps or higher. The term xDSL represents a family of technologies, namely ADSL, SDSL, HDSL, RADSL and VDSL. ADSL and VDSL are the most popular.

How Does DSL Work?
A standard telephone line around the world consists of a pair of copper wires. This pair of copper wires (CAT 1) has far more bandwidth than that required to support voice communication. DSL uses this spare capacity to provide data communications by making use of advanced digital technology.

The key components of DSL are a DSL modem for signal conversion at the customer premises and a multiplexer called DSLAM (digital subscriber line access multiplexer) at the service provider's end. While transmitting, the DSL modem is used to convert the digital signals from a computer into analog so that they can ride over the analog copper wire. While receiving, the DSL modem converts the analog signals back into digital before feeding them into the computer. The DSLAM located at the central office connects multiple DSL users to the Internet service provider (ISP) via a high-speed ATM or frame relay-based backbone (Fig. 5.9).

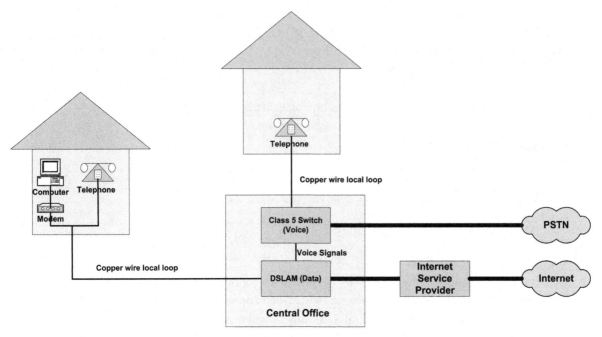

Figure 5.9 - DSL

Family of DSL Technologies
DSL comprises a family of technologies. Different technologies have evolved from developments occurring over the years. The two most prominent are:

- ADSL (asymmetrical DSL) – This is the most widely used DSL technology. ADSL provides higher download speeds of up to 1.5 Mbps, but provides much slower upload speeds of up to 384 Kbps. This asymmetrical speed DSL service is very suitable for homes and small enterprises where a lot of data are downloaded and very little uploaded.
- VDSL (very-high-data-rate DSL) – VDSL is capable of providing download speeds of up to 55 Mbps for distances of up to 1000 ft. The download speed reduces to less than 15 Mbps for a distance of up to 4500 ft. VDSL has gained a lot of attention as service providers are now looking to not only provide traditional voice and high-speed broadband, but also high-bandwidth applications like IPTV and video on demand. Many service providers are implementing fiber to the home (FTTH), but it is prohibitively expensive and time consuming to install fiber to every location. Instead, service providers are running fiber to the node (FTTN) and running VDSL service over existing twisted pair copper wire from the node. This combination of FTTN and VDSL gives service providers a bandwidth of up to 55 Mbps, which is sufficient for all next-generation services.

Advantages of DSL
The biggest advantage of DSL is that it is capable of running over existing age old copper wire infrastructure. This means not much of infrastructure is required to render the services. From user's perspective, the advantage is that the service speed remains constant regardless of the number of users using the service. This is a great advantage in densely populated areas.

Limitations of DSL

The main limitation of DSL is distance. The general rule is - the closer the service location is to central office, the faster the service speeds. The speed is good if the service location is within two to three miles of the central office where the DSLAM is located. To overcome this limitation, service providers are installing DSLAMs at remote terminals (RTs) in order to provide DSL service to locations that are beyond two or three miles from the central office. RTs are mini central offices that are located in neighborhoods with a substantial population but located far from central offices.

5.13.2. FTTH

The existing infrastructure of copper wire in the last mile is placing severe constraints upon the abilities of telecommunication companies to provide additional services like IPTV and host of value-added services like network gaming, gambling, etc. Depending on the distance from the central office or telephone exchange, a consumer can get a connection ranging from 1.6 Mbps to 8 Mbps, a vast majority of whom will only get 1.6 Mbps. Let us take a simple example to show how this capacity is insufficient. A respectable DSL line requires 384 Kbps, a single standard definition TV requires 1.6 Mbps, and a VoIP line requires 64 Kbps, which totals over 2 Mbps. If you add in support for HDTV on multiple TVs (6 Mbps each) and high-speed broadband on computers (3 Mbps or higher) then the bandwidth requirement is even higher. It is expected that an average home would require bandwidth in the order of 36 Mbps. To overcome this constraint, telecommunication companies around the world have started laying optical fiber either all the way to the home or at least up to the node (neighborhood). FTTH (fiber to the home) and FTTN (fiber to the node) is the fastest-growing global broadband technology, with significant deployments in Asia, Europe and North America. FTTP (fiber to the premises) is a term synonymous with FTTH. The generic term is FTTX, where X stands for H, P or N.

FTTX is supported by two deployment architectures:

- Active Ethernet – In this type of architecture, a fiber is run from the central office all the way to the node, and from there a router manages the connection to the premises. This configuration supports very high bandwidth to each user (up to 100 Mbps or higher), but is generally more expensive than other alternatives and requires maintaining active routers and switches that are located in the neighborhood.
- Passive optical network (PON) – In this type of architecture, one fiber is run from the central office or telephone exchange to a passive splitter in a remote terminal. At this point, it is split 32 or 64 times to reach individual homes equipped with optical networking terminals. Unlike active Ethernet, there are no active elements in the neighborhood, which makes it cheaper, but the bandwidth available is much lower and is shared among the users. This is not desirable as it means that the available bandwidth varies depending upon the number of users at any given point in time.

There is a debate raging regarding which of the two technologies is better. Most tier-one providers in the US like Verizon and ATT have favored PON due to cost benefits, but active Ethernet also has its share of proponents.

5.13.3. DSL or FTTX – Which Is Better?

There is no comparison between DSL and FTTX, or for that matter between cable broadband and FTTX. FTTX provides much larger bandwidths (up to a gigabit is possible) compared to DSL (a few megabits) and covers much larger distances as well. DSL broadband is an interim solution so that broadband can be provided on top of existing copper wire infrastructure, whereas FTTX is a true broadband solution capable of supporting next-generation requirements like multiple HDTVs, multiple computers with high-speed connections, and multiple VoIP lines. On the other hand, building infrastructure to support FTTX is far more expensive and it will be a while until it is cost effective to replace the current copper wire in the last mile over which DSL runs. Today, FTTX is the preferred choice for newer communities, while DSL will continue to find use among existing homes where copper wire is already laid. Japan and Korea are leading countries in terms of the number of customers using FTTX broadband.

5.13.4. DSL or Cable Broadband – Which Is Better?

There is constant debate regarding whether DSL or cable is better. The fact of the matter is that neither is inherently better. Theoretically, cable boasts higher speeds than DSL. However, cable broadband is shared among users, so the speed you get at any time depends on how much capacity the other users are taking up. On the other hand, the speed in DSL is distance sensitive; the further you are from central office, the slower the speed you get. They are both equally secure and highly reliable.

Thus, in reality, both DSL and cable broadband are the not very different. The differentiating factor is the service provider.

5.13.5. ISDN

ISDN (integrated services digital network) was the first effort at broadband technology and is a service provided by telephone companies. ISDN also runs on the existing phone line but the technology used is very different from DSL. The biggest disadvantages of ISDN are speed (maximum speeds of only 128 Kbps), cost (more expensive than DSL), and complexity. However, ISDN is a viable alternative where DSL or cable broadband are not available, and is still used in some countries.

5.14. Enterprise Broadband

Broadband plays a very critical role for effective functioning of an enterprise. Enterprises rely on broadband services to support email, internet access, websites, e-learning, corporate LANs, MANs, WANs connecting globally dispersed offices and partners, remote working, voice and video conferencing services. Without these services, any enterprise would virtually come to a halt.

The range of broadband options available to business ranges from simple DS0 (64Kbps) to OC192 (10Gbps) or even higher.

5.14.1. Digital Signal (DSx)

Digital Signal is a term representing the family of connectivity bandwidth available for purchase for the enterprises. The lowest level in the series commonly referred to as DS0 has a bandwidth of 64 Kbps. In addition to DS0, the following DS levels are available:

- DS0 (64 Kbps) - DS0 Service is the ideal point-to-point dedicated digital connection that can be used for voice and data communications. It forms the fundamental base connectivity solution over which the rest of the solutions like DS1 and others are built.

- DS1 (1.54 Mbps or 24 DS0s) - is known as a fundamental building block of many networks. It is a digital, point-to-point, private line service that can be used for voice, data and video traffic. It offers a bandwidth equivalent to that of 24 DS0 lines or 1.544 Mbps.
- DS3 (44.736 Mbps or 672 DS0s) - DS3 Service offers a reliable, all-purpose digital connection for extremely high volume requirements. It transmits video, data, and voice at speeds of 44.736 Mbps in most cases over a fiber optic network.

5.14.2. Optical Carrier (OCx)

Optical Carrier is a term representing the family of high bandwidth connectivity available for purchase for the enterprises. As the name suggests, signals in optical carriers are carried over an optical fiber network. The OC levels are designated as OC-ns and the speed of the level is a n multiple of 51.84 Mbps. For example OC-1 is 51.84 Mbps, while OC-3 is 155.52 Mbps (3*51.84 Mbps).

Higher optical carriers levels like OC-192 (10 Gbps) or OC-768 (39 Gbps) are mainly used to establish the main backbone telecommunication companies operate between their class 4 and international exchanges.

Unlike residential customers, in addition to bandwidth, enterprises also have to choose between different protocols. The primary protocol for LAN is Ethernet, while for MAN and WAN can be ATM, Frame Relay and MPLS that were discussed earlier in section 5.10.

5.15. Wireline Data @ 2010

Data communication began several decades later than voice communication, but today the data traffic on networks exceeds that of voice traffic. In addition, with the advent of IP technology, voice, video and data are all converted into packets of data and transmitted. The future of wireline data networks on the backbone is quite strong. However, there are some serious challengers to wireline data network technologies on the user access side. Wired LANs at homes and enterprises are fast being replaced by wireless LANs. Technologies like Bluetooth and Wi-Fi are quite popular and used extensively. Wireless broadband technologies like WiMax and 3G or higher are posing a serious threat to DSL, cable broadband and FTTX. Thus, it is very likely that the current crop of wired broadband technologies on the access side may be severely challenged by wireless technologies by 2010, but they will continue to exist and provide connectivity where mobility is not required and reliability is very important.

6. Basic Concepts of Cable Communication

Community access television or cable TV was first used in villages across the US in 1948 as a means to receive broadcasted television signals. Since those early days, the industry has grown into a thriving and global telecommunications force. Today, the cable communications industry has over 75% market share in paid television service, not just in the US but also around the world. In addition, the cable industry has expanded into broadband and voice communication services as well.

The total number of worldwide cable subscribers is close to 350 million, with 80 million subscribers in the US alone. However, the fastest growing markets are India and China, accounting for close to 60% of all new cable subscribers during 2007.

6.1. How do Cable TVs work?

Cable companies connect the headend (local distribution center) with many homes using a network of cables (Fig. 6.1) to deliver multiple analog/digital television signals to consumers. The headend receives content through satellite from national networks like CNN, HBO, the local TV stations, and many other sources. Each of the received channels is first converted from different formats to RF signals of different frequencies (frequency division multiplexed). All the signals are combined, amplified, and sent down the shared cable network. The distribution network uses a tree-and-branch topology in which multiple households within a neighborhood share the same cable. The signals going through the network are tapped and sent to users' locations.

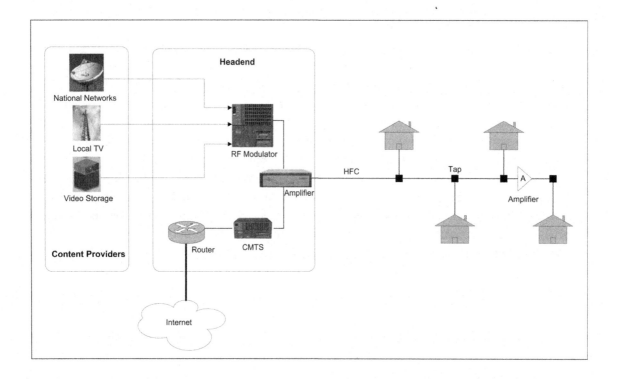

Figure 6.1 – How Does Cable TV Work?

Most cable networks started as one-way communication channels, but today they have rapidly evolved into not only providing video on demand services, but also voice and high-speed broadband connectivity.

6.2. Basic Components

6.2.1. Headend

The headend is the central office equivalent in the cable world, and there is usually one or more in each city depending on the size. It receives content from content providers, modulates it to a format suitable for display on common television, amplifies, and transmits it on the cable distribution network for viewers to access. A simple headend is shown in Fig. 6.1.

Many big cable companies have regional/national headends that receive content from regional and national content providers and redistribute it to local headends using a high capacity optical fiber. A typical headend configuration consists of one super headend per country, one or more headend per city, and many local hubs that are fed from the headend as shown in Fig 6.2.

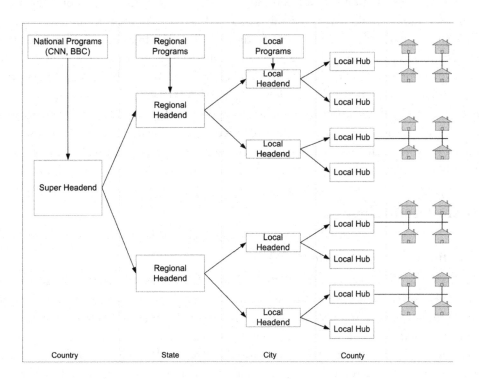

Figure 6.2 - Headend Setup

6.2.2. Set-Top Boxes

Set-top boxes (STBs) are devices that act as an interface between a television and the cable TV network facilitating access to cable services. They perform complex operations like demodulation, decoding, and data decompression so that the incoming signals are suitable for the television to display. In addition, set-top boxes also allow cable operators to manage them remotely (through CAS - conditional access system) so that customers are allowed to view only those channels that they are paying for.

6.2.3. Cable Modem

A cable modem acts as an interface between a computer and the cable network. The cable network is usually analog and computers communicate using digital signals. The cable modem handles conversion from digital to analog and back.

6.2.4. CMTS

A Cable Modem Termination System (CMTS) connects thousands of cable modems to the Internet and is located at the local headend office. CMTS acts as the gateway to the Internet and is capable of receiving and sending signals to and from the cable modem for accessing the Internet. A cable modem sends digital signals, which the CMTS chops into packets and forwards to the local router for transmission across the Internet. On the receiving side, the CMTS receives the packets and modulates the signals for transmission across the cable to the cable modem.

6.2.5. Hybrid Fiber Coaxial (HFC)

The early CATV distribution network was mostly built using simple coaxial cable. The coaxial cable with multiple amplifiers along the way only supported one-way communication. In addition, the amplifiers introduced noise, were prone to failure, and the radio frequency (RF) signals transmitted were susceptible to lightning and other forms of nearby RF signals. To overcome all these problems, the cable companies started using hybrid fiber-coaxial (HFC) cable. It uses fiber-optic cable from the headend to the feeder distribution system and coaxial cable from there to the customer premises. The feeder distribution system contains optical to electrical converters. This arrangement drastically reduces the number of amplifiers required as light signals can travel long distances along fiber-optic cable without requiring any amplification. The addition of fiber increases the available bandwidth and supports two-way operation by simply adding a few upstream amplifiers.

6.3. Basic Terminology

6.3.1. Conditional Access System (CAS)

In a cable system, all channels are transmitted over the cable network and thus are accessible to all customers. However, cable operators have premium channels like HBO, and Sports channels that they want to be accessible to only those customers who pay a premium. CAS technology allows operators to selectively block access to customers who have not subscribed to premium content. It ensures that only those users who are authorized can watch the program. Without CAS, cable operators would not have been able to offer premium channels, pay per views, support password access control, and anti-copy control.

CAS uses scrambling and encryption techniques. The video and audio streams at the operator's end are scrambled and a key is transmitted to the receiving set-top box. The set-top box is capable of descrambling the signals and using the keys to make the video and audio available.

6.3.2. Digital Rights Management (DRM)

DRM is a mechanism used to protect the rights of content producers, protecting them from unauthorized use of their copyrighted material like music, films, text, or any other form of content. It represents a family of technologies that defines access to copyrighted material. In essence, it removes usage control from the person in possession of the digital content and puts it in the hands of a computer program, thereby preventing users from making illegal copies.

DRM is one of the most controversial topics in content management, with fierce proponents and opponents. Opponents see DRM methods as overly restrictive, even though DRM is an effective means to solve a legitimate problem. Proponents argue that the music and movie industries lose billions of dollars in sales due to illegal sharing through Internet sites like Napster and Kaaza.

6.3.3. Video on Demand (VoD)

Consumers spend billions of dollars on renting content like movies, and probably millions more in late fees. Renting movies is expensive and a lot of work. VoD is a mechanism that allows customers to order a movie with just a few clicks on the remote, and the movie or any other content like a boxing match will be rendered directly over the TV set. The content is provided by storing content on a server and streaming it to the viewer upon their request. The user can stop, start, fast forward or rewind the video, just as if they were watching a recorded movie on their DVD player. Users can watch the content as many times as they want over a fixed period, or even buy it.

This is an extremely attractive proposition for renting movies as the customer doesn't have to leave the house to rent a movie, nor does he have to worry about late return penalties. VoD is also an extremely attractive revenue opportunity for cable operators because of the strong demand among existing residential subscribers. In addition, VoD is useful in many other applications like e-learning, training, marketing, entertainment, and other areas where the user would like to view the files at their convenience. It is a key revenue-generating service for service providers and has the potential to increase the average revenue per user (ARPU) by as much as 25%.

6.3.4. Personal Video Recorder (PVR)

PVR is a device capable of recording a video stream while it is being broadcast so that a user can watch that show at any other time. PVR is not a new concept and is commonly found in the US from vendors like TiVO. It allows the user to schedule a TV recording, view and pause live TV, acts as a media center to watch movies, listen to music, view pictures, listen to FM radio, and record multiple shows at the same time.

6.4. TV Technologies

The quality of viewing TV is largely dependent on the technology used. There are two prominent TV technologies in use today:

- Analog
- Digital

6.4.1. Analog TV

Analog TV was the first TV technology, and has been in operation for over 50 years. However, the quality of viewing is not very good as compared to digital technology. The signal is easily distorted by terrain, tall buildings, or electromagnetic sources.
There are two main analog TV standards: NTSC in the US, and PAL in Europe and many other parts of the world.

6.4.2. Digital TV (DTV)

Digital TV delivers a picture that is far clearer and sharper than analog TV. The reason for better performance is because the signals are digital that are even if distorted can easily be filtered to provide movie-quality picture and sound. In addition, digital TV allows operators to provide far more channels, and interactive capabilities like voting as well.

There are two types of digital TV: standard definition and high definition. The difference is in the number of lines they use in each frame. The number of lines determines the amount of detail in the image - the higher the number, the better the picture quality.

- Standard definition TV (SDTV) – Digital televisions capable of receiving either 520 or 625 lines.
- High definition TV (HDTV) – Digital televisions capable of receiving either 720 or 1080 lines. HDTV offers the better audio and picture quality, with clear, sharp images and digital quality sound.

Apart from the technicalities, and the difference in picture quality, the other difference between SDTV and HDTV is that the signal on SDTV is more compressed than that of HDTV. A channel in SDTV requires far less bandwidth (about 1/5th) as compared HDTV. Thus, providers can stream more channels of SDTV and offer variety.

6.5. Standards Bodies

There are several organizations involved in setting standards for various aspects of television services, some of which are as follows:

6.5.1. Digital Video Broadcasting (DVB)

DVB is an industry-led consortium of over 300 broadcasters, manufacturers, network operators, software developers, and regulatory bodies. They are present in over 35 countries around the world, and are committed to designing global standards for the delivery of digital television and data services. They covers all aspects of digital television from transmission through interfacing, conditional access and interactivity for digital video, audio and data.

DVB's main transmission standards are DVB-S for satellite, DVB-C for cable, and DVB-T for terrestrial. These standards are used extensively around the world.

DVB has dedicated itself to defining standards that are open so that compliant systems will be able to work together. Designed with the maximum amount of commonality and based on the MPEG2 coding system, DVB signals may be effortlessly carried from one medium to another, which is a frequent need in today's complex signal distribution environment.

6.5.2. CableLabs

CableLabs is a research and development consortium of cable television system operators representing North and South America. CableLabs plans and funds research and development projects that will help cable companies take advantage of opportunities and meet challenges in the cable television industry. It serves the cable television industry by researching and identifying new broadband technologies, authoring specifications, certifying products, enabling interoperability among different cable systems, defining advanced services, and helping cable operators deploy innovative broadband technologies.

The main standards defined by CableLabs are PacketCable 2.0, Open Cable, and DOCSIS.

6.6. Video Distribution Standards

The analog TV world is divided into three incompatible video signal distribution standards - NTSC, PAL, and SECAM. Depending on where you live, the technology used for signal transmission by the cable companies is different. NTSC is the standard in North America, South America and Japan. PAL is standard in Europe. Luckily, analog TV technology is fading out and the digital world is promising uniformity in MPEG standard. Note – these standards are also used by broadcast TV companies.

6.6.1. NTSC

NTSC, or National Television Standards Committee, is a video signal distribution standard for analog television developed and used extensively in the US and Japan. A composite video (containing video and audio) signal requires each frame to have 525 lines, with a screen refresh rate of 60 half-frames interlaced per second. In the NTSC system, each channel occupies a total bandwidth of 4.2 MHz.

6.6.2. PAL

PAL, or phase alternating line, is the dominant video signal distribution standard for analog television, developed and used extensively in Europe and many parts of the world. A composite video (containing video and audio) signal requires each frame to have 625 lines at 50 half-frames interlaced per second. In PAL, each channel occupies 5 MHz.

6.6.3. SECAM

SÉCAM (Sequential Color with Memory), is an analog color television system first used in France and is used today in Russia and couple other European and African countries. It is, historically, the first European color television standard.

6.6.4. MPEG

MPEG, or Moving Picture Experts Group, is the most popular video signal distribution standard for digital television and computers. It is an open source committee in charge of the development of standards for the coded representation of digital audio and video. Established in 1988, the group produced the first standard called MPEG1, which is still used in video CD and MP3. MPEG1 defines standards for compressing and combining the audio and video signal into one single stream, which makes it particularly suitable for digital storage or transmission. The MPEG2 standard, approved in 1994, was developed to support digital television and DVD (digital versatile disk!) and represented significant enhancements over MPEG1. MPEG4, approved in 1998–99, has been selected by several industries for next-generation communications. It is also being utilized as standard for Mobile TV and IPTV.

In addition to these, MPEG group has also released two standards – MPEG 7 and MPEG 21. These are not standards used for compression like MPEG1, 2, and 4, but are meant to describe the features like title, synopsis, year of release, director & actor information. This kind of metadata information is very useful to provide a simple, flexible, interoperable solution to the address issues around content distribution like indexing, searching and retrieving multimedia resources. In addition, MPEG-21 also contains information about rights management about content so that it can be used by a conditional access system to enforce them.

6.7. Cable Technologies and Standards

In addition to the digital and analog signal transmission technologies, Cable companies also have several standards for broadband, interactive and multimedia services over TV.

6.7.1. DOCSIS

DOCSIS, or Data over Cable Service Interface Specifications, is a standard developed by CableLabs for providing Internet data services to users. It defines interface standards for cable modems and other supporting equipment required for Internet access. It specifies a downstream data communication rates between 27 and 36 Mbps, and upstream rates between 320 Kbps and 10 Mbps.

6.7.2. DOCSIS 3.0

The DOCSIS 3.0 specification first came to light in August, 2006 when CableLabs outlined the methodology for downstream and upstream channel bonding, along with other features such as IPv6, IP multicasting and AES encryption. By early 2008, many providers like Time Warner Cable are aggressively implementing this standard. It can achieve downstream speeds of up to 160 Mbps and upstream speeds can be in the order of 120 Mbps. In addition to high upstream and downstream speeds, DOCSIS 3.0 being an IP based standard will allow the cable companies to deploy interactive video services like video on demand, network gaming, etc.

6.7.3. PacketCable 2.0

PacketCable is a CableLabs-led initiative that is aimed at developing interoperable interface specifications for delivering advanced, real-time multimedia services over two-way cable plant. Built on top of the industry's highly successful cable modem infrastructure, PacketCable networks use Internet protocol (IP) technology to enable a wide range of multimedia services, such as IP telephony, multimedia conferencing, interactive gaming, and general multimedia applications. Working with CableLabs member companies and technology suppliers, the PacketCable project addresses issues such as device interoperability and product compliance with PacketCable specifications.

6.7.4. OpenCable

OpenCable is a set of hardware and software specifications under development in the US by Cable-Labs, with the goal of helping the cable industry deploy interactive services over cable. Just like several other CableLabs projects, including DOCSIS and PacketCable, OpenCable provides a set of industry specifications.

The OpenCable project has two key components - a hardware specification and a software specification. The hardware specifications describe both one-way and two-way digital cable-ready "host" devices that are interoperable with cable systems throughout the US. The software specification of the OpenCable project, called the OpenCable application platform (OCAP), solves the problem of proprietary operating system software, thereby creating a common platform for interactive television applications and services.

6.7.5. Multimedia Home Platform (MHP)

MHP is the open middleware system designed by the DVB project. MHP defines a generic interface between interactive digital applications and the terminals on which those applications execute. The standard enables digital content providers to address all types of terminals ranging from low to high-end set-top boxes, digital TVs, and multimedia PCs. With MHP, DVB extends its successful open standards for broadcast and interactive services in all transmission networks, including satellite, cable, and wireless systems.

MHP has been formally adopted by the European Telecommunications Standards Institute (ETSI).

6.8. Cable Broadband

Cable broadband is the broadband solution offered by cable TV providers. The popularity of cable broadband arises from the fact that the existing cable that runs to a service location carrying TV signals can also accommodate high-speed data communication.

6.8.1. How does Cable Broadband work?

The CATV network was initially built to be a one-way communication network, but today's communication applications like voice and Internet access require the network to support two-way communication. Thus, during the early 1990s, cable companies went through a massive upgrade of their networks, which is enabling them to support two-way voice and Internet access services. Along with multiple video channels, one or more channels are set aside for carrying data for Internet access. Users can connect their computers using a cable modem, which communicates with the CMTS (cable modem termination system) located at the headend, as shown in Fig. 6.1. The CMTS is capable of handling thousands of cable modems and acts as the gateway to the Internet.

6.8.2. Advantages of Cable Broadband

Cable broadband is a very attractive and popular broadband access technology. The key advantages are:

- Deployment – The existing cable can be easily reused for broadband with very little additional equipment.
- Cost – Cable broadband requires less infrastructure than DSL, hence is comparatively cheaper.

6.8.3. Limitations of Cable Broadband

Theoretically, the coaxial cable from a cable company is capable of providing much higher speeds compared to DSL, but in reality, it is not. Unlike DSL, which provides a dedicated channel for broadband, cable broadband is a shared broadband service. Cable companies connect consumers by forming local area networks (Fig. 6.1). The speed of service is directly dependent upon the number of users accessing the service at any given time. This means that speeds decrease when there are too many people using the same cable.

6.9. Cable over IP

Internet Protocol or IP technology (chapter 9) is revolutionizing the field of telecommunications. It provides enormous cost savings, converged services with rich features and host of other benefits. Cable companies around the world are extensively using this technology to provide voice (Voice over IP or VoIP) and broadband services already and are on the way to use it to provide television services as well.

6.9.1. VoIP Over Cable

There is no denying that triple play bundles (voice, video and data) are one of the best ways to keep customers. Cable companies have a strong advantage over other telecommunication companies as they already provide television and broadband services and are now capable of providing voice services using IP technology. Cable companies are even seeking tie-ups with wireless phone companies so that they can offer a quadruple play bundle that also includes a mobile phone service. Many companies are investing heavily in offering the triple play bundle as they are fast realizing it is the best way to retain customers and improve satisfaction.

The demand for voice service from cable companies is evident from the strong demand seen in the US. There were around half a million VoIP subscribers by the end of 2004, which increased to 4.7 million at the end of 2007, representing a 900% increase in two years. A sizeable chunk of these belongs to cable companies.

6.9.2. IPTV over Cable

Most people view IPTV (TV based on Internet protocol, explained in chapter 9) as a wireline centric business. While it is true that traditional wireline companies are aggressively looking to deliver TV and video over their own IP infrastructure, but cable companies are not far behind.

The marketplace is demanding that cable operators deploy more bandwidth-intensive products like HDTV, video on demand (VoD), and potential future services like network-based PVR (personal video recorder), network gaming, video telephony, and a host of interactive applications. These new demands cannot be easily met with the current infrastructure and technology. To overcome these challenges, cable companies have begun seriously exploring the possibilities of using Internet protocol (IP) and MPEG4 as the way to provide services.

6.10. Cable @ 2010

The one thing that every cable company is seriously considering is migrating to an all-IP network, which would provide enormous benefits. Such a move would enable them to provide a lot more channels using existing cable, and provide truly convergent and highly interactive services that the marketplace is demanding. In addition, they can take advantage of the cost benefits of a unified network where video could run on the same IP infrastructure used for broadband and VoIP. This move will also lower operational and equipment costs, and provides the ability to migrate to standards-based security versus the more expensive conditional access technologies. It is very likely that by 2010, many cable companies will have started migrating to an all-IP network.

7. Basic Concepts of Wireless Voice Communication

Wireless communications uses radio waves (part of the microwave spectrum) rather than some form of wire or fiber to carry the signal over part of or the entire communication path.

The wireless communications industry can be primarily divided into two categories:

- Mobile wireless (Cellular based): cell phones
- Fixed wireless (Network based): Wireless LANs, Wi-Fi, Wimax, etc.

In both voice and data business, wireless mode of communication is fast overtaking wireline mode of communication. Today, there are more cell phone users around the world than wireline phones and people are extensively using technologies like WLAN (wireless LAN) and Wi-Fi for data communications. According to the International Telecommunications Union (ITU), the current wireline voice subscription around the world stands at 1.85 billion, while cell phone subscriptions top 2.7 billion and continue to show robust growth in all continents. By the end of 2007, the global mobile business was worth more than US$550 billion and growing very rapidly. China is the world's largest mobile phone market with over 260 million subscribers, while the US comes in second with 153 million. India is the world's fastest growing market with a CAGR of at least 20% per year until 2010.

The future for wireless communication is very bright. Explosive growth is expected to come from all sectors (voice, video and data) around the world. 3G and WiMax are touted to be the next-generation technology for wireless broadband services that will eventually make a wireless triple play of voice, video and data possible. Meanwhile, 4G is waiting in the wings with promises of even higher bandwidth. In addition, there is considerable interest in wireless VoIP technology by enterprises around the world.

7.1. Cell Phones

7.1.1. How Do Cell Phones Work?

When a user places a call, the cell phone connects to the nearest cell tower (as shown in Fig. 7.1), also called a cell site or base station, using high frequency radio signals. Each cell phone uses a pair of frequencies to communicate with the base station, one for communication from the tower to the cell phone, and another for communication from the cell phone to the tower. Each cell tower contains a transmitter and a receiver, along with equipment to power and control the base station. Every base station is connected to a central office equivalent called a mobile telephone switching office (MTSO) through high capacity fiber or microwaves. There is usually one MTSO per carrier per city.

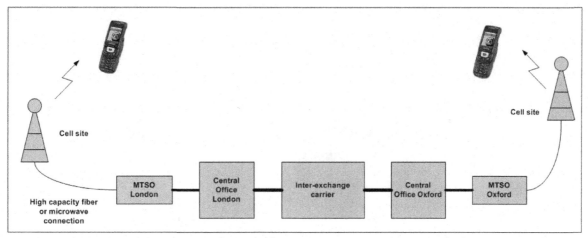

Figure 7.1 - How Do Cell Phones Work?

7.2. Basic Components

7.2.1. Mobile Telephone Switching Office (MTSO)

A MTSO is the wireless equivalent of a central office or telephone exchange. Multiple base stations or cell towers are connected to each MTSO (Fig. 7.1) through high-capacity trunk lines. Just like a central office, an MTSO also contains switches that are used to switch the call to the receiver. If the receiver is another cell phone connected to a base station within the same MTSO, the call is switched directly. If the receiver is not within the same MTSO then the call is handed over to an inter-exchange carrier (if it is long-distance) or a tandem exchange/central office (if it is local), depending on where the receiver is located.

The process of connecting a wireless phone to its destination is quite complicated compared to connecting a wireline phone. It is quite possible that the user is moving from cell to cell, the user can call another cell or a wireline phone, and the user can not only conduct voice communication, but also browse the Internet. All this complexity is smartly handled by the MTSO, which makes sure that the cell phone is connected to the nearest cell site that gives the best reception, seamlessly manages hand-offs from one cell site to another as the user moves, and serves as a conduit between the cell phone and the PSTN (via central office). Apart from managing the logistics of connecting calls, the MTSO is also capable of provisioning and activating cell phones and generating billing events.

7.2.2. Cell Sites

A cell site is the physical location where the tower containing the antenna, transmitter, receiver and other equipments is located. Each cell site contains multiple transmitters and receivers, power sources, antennas, and equipment to transmit the signals to MTSO. The transmitters located in a cell site are low powered and their reach does not go beyond the boundaries of the cell. A cell site located at the node (refer Fig. 7.2) usually covers two to three cells. Each site is usually shared by multiple carriers, each providing their own equipment but sharing the same tower.

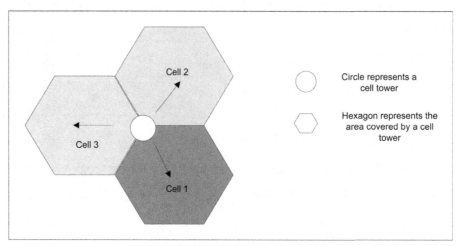

Figure 7.2 - Cell Site

7.2.3. Cell Phone

A basic cell phone is just a simple two-way radio. The basic components of a cell phone are a transmitter, receiver, antenna, and power source. Low power requirements make all wireless devices like cell phones small and handy.

Modern cell phones (e.g., iPhone, Nokia n95) are much more complicated, and capable of providing many services beyond just voice communication. They have powerful processors and rival computers in terms of capabilities. Apart from handling voice communications, they provide almost all the capabilities of a computer, like browsing, email, word processing, spreadsheets, storage, music, games, and video conferencing.

7.3. Basic Concepts

7.3.1. Cell

The most amazing thing about the cellular phone network is the cell itself. A cell is defined as the area covered by a cellular tower (as shown in Fig. 7.2), and all cell phones within that area are served by that tower. The area covered depends on the power of the transmitter and the frequency used. A cell phone communicates with the tower servicing the cell that it is physically located in. A cell is considered to be hexagonal because a hexagon best represents the idea of a few towers totally covering a geographic area. A circle leaves gaps and there are complications in representing an area with triangle or squares. The size of each cell depends on the terrain, the power and frequency of the equipment in the cell tower. It is very likely that a given spot may be served by more than one tower, but the cellular device connects to the one whose signal is the strongest.

7.3.2. Frequency Reuse

The main constraint in wireless communications is the range of frequencies available for communication. A federal authority in each country allocates a fixed frequency range to each carrier within a geographic area. The range of frequencies distributed among carriers in different countries is as follows:
- 850 MHz (US)
- 900 MHz (Europe/Asia)

- 1800 MHz (Europe/Asia)
- 1900 MHz (US)

Federal authorities in many countries have also started auctioning 400 MHz for use in rural areas.

The frequency band allocated to each carrier gives it 800 channels, which is sufficient to serve 395 conversations at a time as each customer needs two channels, one for communication from tower to cell phone and one for cell phone to tower. This is a very small number, probably not sufficient to even support a small area of a town. To overcome this deficiency, each carrier divides the geographic area in which it provides service into small hexagonal cells. A group of several adjoining cells formed by a tower, called cluster can reuse the entire range of frequencies allocated to the service provider. For example, each cell in a seven-cell cluster (Fig. 7.3) receives one seventh of the frequency, or 115 channels. Similar numbered cells receive the same frequency range. For example, cell 1 in each cluster always gets channels 1 to 115 and cell 2 in each cluster always gets channels 116 to 230. The adjoining cells are never of the same number, which assures that channels are not mixed up.

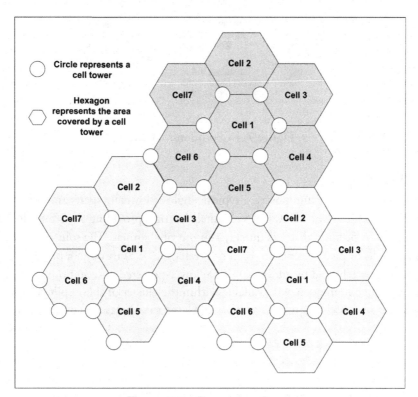

Figure 7.3 - Frequency Reuse

7.3.3. Frequency Band

The band of frequency the wireless provider uses play a very important role in terms of the geographic reach and the number of subscribers served by each tower. As mentioned earlier, each carrier is given 800 frequencies in one of the four bands - 850/900/1800/1900 MHz. The general rule is - the lower the frequency, the larger the coverage area, but lower the number of communication channels (Fig. 7.4). For example, a tower operating on 850 MHz can reach farther than a tower operating on 1800 MHz, but is capable of serving a smaller number of customers. Therefore, a smaller frequency band is useful in rural areas where the user density is low and the user population is spread out. A higher frequency band is useful in densely populated areas like downtown where there are lots more users within a small geographic area.

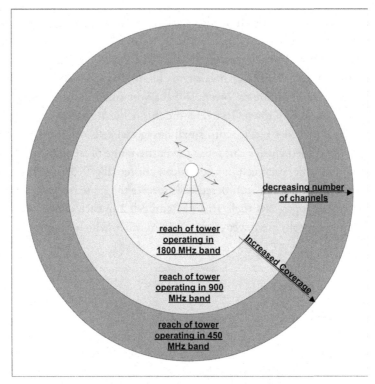

Figure 7.4 - Frequency Band

7.3.4. Cell Splitting

Depending on the frequency and power, a typical single cell usually spans around 10 square miles at full power. Since each cell only allows 115 channels, carriers providing service in densely populated areas like downtown business districts quickly run out of channels. The solution to this problem is to create micro cells or pico cells within each cell (refer Fig. 7.5) through a process called cell splitting. Cell splitting is achieved by adding multiple lower-powered base stations that cover a smaller area. The entire range of frequency is available within the cluster of these split cells. Carriers split cells several times in densely populated areas, while large rural areas are covered by a single cell.

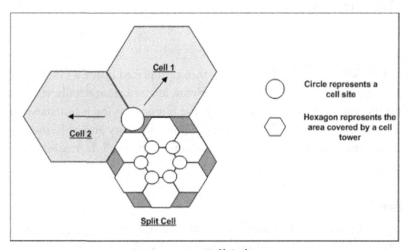

Figure 7.5 - Cell Splitting

7.3.5. Handoff

A cell phone user driving along the highway might drive for 50 miles before hanging up. Obviously, a single cell cannot cover a distance of 50 miles, so the caller has to be handed off from one cell to an adjacent cell. The handoff process, managed by MTSOs, is complicated by the fact that adjacent cells use different frequencies. As the user starts moving toward the edge of a cell or away from a cell site, an MTSO scans for the strongest available signal. The call is then handed over to the cell site that provides the strongest signal strength and thus the user can continue conversation without being disconnected.

7.3.6. Roaming

Every service provider has a fixed geographical area (home calling area) where it provides service. However, you can easily use a cell phone outside of the provider's area as long as there is another provider using similar technology – GSM/CDMA and frequency band (850/900/1800/1900 MHz) – in that area. Roaming is the term used to describe the ability of a cell phone or other wireless device to use the facilities of other service providers while outside the home calling area. This can sometimes also occur within the home calling area, with at least two service providers usually licensed to provide service within each geographic area. If the signal from the carrier the user is subscribed to is weak or overloaded, the wireless device switches to using the other service provider's facilities.

The advantage of roaming is that service is provided seamlessly, instead of dropping or blocking the call when service cannot be provided to the user by the provider to whom they are subscribed. Technology has enabled worldwide roaming, which has greatly enhanced the popularity of cell phones. The only disadvantage of roaming is that the charges are exorbitant. The worldwide roaming market is calculated to be worth around US$50 billion, 40% of which comes from the Asia-Pacific region.

7.4. Basic Terminology

7.4.1. SIM

A Subscriber Identity Module (SIM) is a small, thin card placed inside a GSM phone. It carries the telephone number of the subscriber, encoded network identification details, the user's PIN, and other user data such as the phone book. It allows the phone carrying it to make and receive calls, and stores personal information such as the user's phone directory and SMS messages. A user's SIM card can be moved from phone to phone as it contains all the key information required to activate the phone.

7.4.2. Quad-Band

In any country, GSM and CDMA providers are allotted two out of the four bands of frequency ranges (850/900/1800/1990 MHz) for providing service within a geographic area. Quad-band phones are capable of working in all four bands. Devices having ability to operate under more than one frequency band is useful to enable roaming between different countries that peg the allowed transmission frequency at different values, or sometimes to allow better coverage in the same country.

7.4.3. PTT

Push to talk (PTT) is a simple walkie-talkie type of communication. A user can dial any number and follow the press–hold–talk over routine while talking, and release it for the other party to speak. Users can go to their address book and establish a PTT connection with any of their friends. The advantage of PTT is that no calling minutes are used up, but there is usually a monthly fee.

7.4.4. FMC

Fixed-mobile convergence (FMC) is about the integration of wireline and wireless technologies so that wireless devices can seamlessly utilize the wireline infrastructure. A simple example of FMC is when a cell phone switches over to cheaper wireline infrastructure while at home or office rather than using the expensive wireless spectrum.

7.4.5. WAP

Wireless Application Protocol (WAP) is a set of communication protocol standards to make accessing online services from a mobile phone simple. WAP was conceived by four companies: Ericsson, Motorola, Nokia, and Unwired Planet. It allows cell phones to access web pages and other advanced services over Internet.

7.5. Cell Phone Technologies

There are two key technologies used for cell phone based voice communication:

- GSM - Global System for Mobile Communications
- CDMA - Code Division Multiple Access

Both GSM and CDMA are families of technologies that support voice and data communication (Chapter 8).

7.5.1. GSM

GSM stands for Global System for Mobile communications and is a very popular wireless technology. The technology, first demonstrated in Geneva in 1991, has picked up very rapidly since then. At the time of writing (late 2007), the number of GSM subscribers around the world stood at a staggering 2.6 billion, which accounts to a market share of 82% among different wireless technologies. This also means that one in three people around the world use GSM-based technologies for wireless communications, which is quite an achievement.

The biggest strength of GSM are as follows:

- Open non-proprietary system - Being open source has allowed GSM to enhance its capabilities very rapidly as many companies and individuals contribute to the development of the technology
- SIM Cards - SIM cards allow users to store personal information like their contact phone list, personal phone settings, and other important information. A user can buy a new phone and still use the same SIM card
- Interoperability and Roaming - GSM users can use their phone in almost every country with the same telephone number. International roaming and a standardized service anywhere in the world have proven to be the biggest attractions of GSM technology
- Good consistent quality – GSM uses time division multiplexing technology. Multiple users use the entire frequency band and transmit for a short time in round robin fashion. Thus, at any given time, only a fixed number of users are allowed to use the service that ensures good consistent quality. However, if the number of users increases then GSM has frequency hopping capability, whereby users are seamlessly switched from the 800 MHz to other frequency band.

7.5.2. CDMA

CDMA stands for code division multiple access. It was extensively used by the allied troops during the Second World War, but did not find commercial acceptance until 1995. Qualcomm Communications from the US first developed chips that implemented CDMA technology, and today holds most of the patents, making this technology proprietary. At the time of writing (late 2007), the number of CDMA subscribers around the world stands at 421 million, with 44% of the customer base in Asia, 35% in the US, and only 2.1% in Europe and the Middle East.

CDMA technology is neither based on principles of time division multiplexing nor frequency division multiplexing. CDMA uses a unique code to establish channels of communication between the wireless device and the tower. The advantage of this is that the full spectrum of frequencies is available for each channel. Establishing a unique code is equivalent to making each user talk a different spectral language, which is decoded at the tower.

7.5.3. CDMA or GSM – Which Is Better?

The battle between CDMA and GSM technologies around the world is quite intense. In terms of market share GSM is the clear leader, with 2.6 billion subscribers compared to 421 million for CDMA, but the story of which is better is not reflected by these numbers. To answer this, the two technologies are compared based on a few important parameters.

Capacity

GSM can serve a fixed number of users per cell, while there is no such limitation in CDMA. This is a great advantage for CDMA providers as a seemingly endless number of users can be supported at any time. However, the problem is that the quality of voice deteriorates as the number of users increases; this is referred to as channel pollution and is a cause of concern among many CDMA users, especially those living in densely populated areas. On the other hand, GSM supports only a finite number of users per cell. This means that voice quality remains consistent and good, but it is difficult to achieve connectivity in densely populated areas.

In addition, CDMA cell sites are capable of supporting a much wider customer base than GSM. A typical site covers up to 69 miles, compared to GSM's 21 miles.

Roaming

International roaming is a clear disadvantage for CDMA as relatively few countries (little more than 30) provide such a service using CDMA. Meanwhile, GSM is used in almost all countries. Therefore a person with a quad-band GSM phone can get phone service anywhere in the world without even changing the phone number. This is extremely beneficial to frequent travelers.

Handoff

The handoff in CDMA is much smoother than GSM. As every tower uses the same frequency, the CDMA handset only needs to generate a new code to start communicating with the other tower. This allows a handset to communicate with multiple nearby towers during handoff and switch to the most capable. This simultaneous communication with multiple towers during handoff allows the handset to switch between towers without the user ever knowing. This is referred to as soft handoff technique.

In GSM, the handset constantly compares the strength of the signal from the tower it is currently communicating with to the strength of the signal from towers nearby. This information is transmitted to the MTSO, which determines when to switch to another tower. When the time for switching from one tower to another comes, the call is literally dropped for a moment and switched over to the other tower before being connected again. This handoff is seamless most of the time, but sometimes calls are dropped. This is referred to as hard handoff technique.

Talk Time

GSM has a clear edge over CDMA in this category, with GSM devices having a much longer talk time and standby time. This is because CDMA transmission between the tower and the device is always on, while in GSM transmission occurs only when there is something to transmit.

As you can see from the above discussion there is no clear winner in terms of technology. Each has its own advantages and disadvantages, and they both seem to provide similar features and capabilities. According to Nokia, "this discussion is not about technology anymore, but about market" and this is the best way to describe the war between these two cell phone technologies. Considerable research has been carried out to discover the criteria that customers use when buying a cell phone, and it has been consistently proven that regardless of the technology, customers choose a provider based on the reliability and quality of the service provided and not the technology.

7.6. Technology Standards Evolution

The standards for wireless technology are addressed using 1G, 2G, 3G and 4G, where G stands for generation. These standards are defined by standards bodies like 3GPP, 3GPP2, ETSI, IMT, and others. Each generation of technology has capabilities better than the previous generation in terms of voice quality, higher data transmission rate for Web browsing and emailing, and better overall security. The standards for each generation are defined by International Telecommunication Union (ITU - a UN body). Both GSM and CDMA closely follow and implement them.

7.6.1. 1G

The first generation of wireless technology began in the 1980s. The 1G network provided an analog voice service, but no data service. It was the first wireless standard and as such did not use the spectrum efficiently. It had many problems with quality and security, and handoff was very rudimentary.

1G was not really the first technology for wireless communications; there were several mobile radio networks in use before its development. However, the major difference between 1G technology and its predecessors was that 1G was based on the concept of cells. The most successful 1G standard was advanced mobile phone service (AMPS), while there were some other standards like Nordic Mobile Telephone (NMT) and Total Access Communications System (TACS).

7.6.2. 2G

Unlike 1G, second-generation or 2G wireless technology is digital. The use of digital technology allows multiple channels to share the spectrum by frequency or time division multiplexing the channel, thereby increasing spectral efficiency. 2G technology also introduced the concept of splitting cells into micro or pico cells, which led to an enormous increase in capacity.

GSM and CDMA are the two most widely used 2G technologies. 2G has good support for voice communications but has limited data communications capability. The top speed achieved by 2G technologies is up to 14.4 Kbps. 2G phones will be succeeded over the next couple of years by 2.5G and 3G phones.

7.6.3. 2.5G

2.5G is the most widely implemented technology around the world today (2008), and is essentially an intermediate standard between 2G and 3G phones. 2.5G technology involves a software enhancement over existing 2G technologies, making data communication packet-based. 2.5G also has a much higher bandwidth speed of up to 56 Kbps. A new wireless standard, Enhanced Data GSM Environment (EDGE), has been developed to increase the bandwidth to a peak rate of 384 Kbps, allowing GSM operators to offer high-speed data communication services.

7.6.4. 3G

The third generation (3G) of wireless technologies promises better voice quality, much higher data communication speeds, and a host of multimedia services. The ITU (International Telecommunication Union – a UN body) mandates that 3G networks, among other capabilities, deliver improved system capacity and spectrum efficiency over the 2G systems and support data services at minimum transmission rates of 144 kbps in mobile (outdoor) and 2 Mbps in fixed (indoor) environments.

Currently only CDMA's CDMA2000 and GSM's WCDMA qualify as 3G standard. Because 3G networks use different transmission frequencies and require different infrastructure from 2G and 2.5G networks, operators and carriers are incurring significant expenses (in the billions of US dollars worldwide) in order to buy spectrum licenses and develop their 3G infrastructure.

7.6.5. 4G

4G is the next generation of standards and is not yet fully developed. In the simplest terms, 4G is the next generation of wireless networks, which will replace 3G networks at some point in the future. 4G is expected to address the limitations and problems of 3G in terms of performance and throughput. For example, 3G is not capable of meeting the needs of future high-performance applications like multimedia, full-motion video, or wireless teleconferencing.

4G standards are likely to recommend an all-IP network and offer peak data communication rates of up to 100 Mbps while roaming and 1 Gbps for fixed. It plans to utilize IP in its fullest form with converged voice and data capability.

7.7. Standards Bodies

Several standards bodies are setting the standards for voice, video and data communications over wireless networks, some of which are described below.

7.7.1. 3GPP

The 3rd Generation Partnership Project (3GPP) is a group formed by standards bodies like ARIB (Japan), CCSA (China), ETSI (Europe), ATIS (USA), TTA (Korea), and TTC (Japan). The scope of 3GPP is to provide technical specifications for 3G GSM technologies for voice and data. The specification covers all GSM technologies like GPRS, EDGE, WCDMA, and UMTS.

7.7.2. CDG

The CDMA Development Group (CDG) is an international consortium of companies who have joined to lead the adoption and evolution of 3G CDMA wireless systems around the world. The primary responsibility of CDG is to accelerate the definition of requirements for new CDMA features, services and applications, and define the evolution path for current and next-generation CDMA systems.

7.7.3. 3GPP2

Just like 3GPP, 3GPP2 consists of the same group of standards bodies: ARIB, CCSA, ATIS, TTA and TTC. While 3GPP defines standards for GSM-based 3G WCDMA technology, 3GPP2 defines standards for CDMA-based 3G CDMA2000 technology.

7.7.4. IMT-2000

International Mobile Telecommunications 2000 (IMT-2000) is an ITU (International Telecommunication Union – a UN body) based organization that sets the global standard for 3G wireless communications. It provides a framework for worldwide wireless access by linking the diverse systems of terrestrial and/or satellite-based networks, and exploits the potential synergy between digital mobile telecommunications technologies and systems for fixed and mobile wireless access systems.

The defined standards are for frequency spectrum and technical specifications for radio and network components, tariffs and billing. They also provide technical assistance and studies on regulatory and policy aspects to operators around the world.

7.8. Wireless Voice @ 2010

Wireless voice has come a long way, and is set for massive growth in the coming years. Research by various organizations predicts that the number of wireless users worldwide could easily reach 3.5 billion by 2010, which would represent more than half the world's population. According to the World Bank, over two thirds of the world's population is already within the reach of a wireless network, and with voice as the mainstay mode of communication, this sector is bound to see tremendous growth for years to come.

Most developed countries have already reached saturation levels in terms of mobile subscriptions. The growth in wireless voice is expected to come from Asia Pacific, Africa, Latin America, and the Middle East. These markets are bound to offer exciting growth prospects for mobile operators, provided they formulate cost-effective and reliable services for their customers.

In the coming years, most of the development in wireless voice communication will be around implementing 3G and 4G technologies. However, the focus over the next couple of years will be to manufacture low cost handset, which can significantly contribute in making wireless services affordable in developing countries.

Despite tremendous growth and the popularity of wireless voice service, there are still concerns regarding quality, coverage and cost. In each of these areas, wireless communication is no-where close to providing the same customer experience as wireline voice service. However, with new technology and access methodology, the gap is closing fast. Many federal authorities are allocating additional spectrum to providers in order to ease congestion and improve quality. The day when all communication devices are wireless is not very far away…

8.   Basic Concepts of Wireless Data Communication

Just like wireless voice, the wireless data market has been registering phenomenal growth over the last few years. The market is led by 2.5G, 3G and Wi-Fi (802.11x) technologies. By late 2007, over 40% of companies in North America and Europe had already wireless-enabled their LANs. Last year, the number of worldwide Wi-Fi hotspots grew 87%, from 53,779 in 93 countries to 100,355 hotspots in 115 countries. The number of Wi-Fi hotspot users worldwide is growing at the rate of over 200% every year. By the end of 2007, over 70% of households in North America and Europe had already wireless-enabled their homes.

The future for wireless is very bright, with continued explosive growth coming from voice, video and data markets around the world. 3G and WiMax are touted to be the next generation technologies for wireless broadband services, which will eventually make a wireless triple play containing voice, video and data possible. There is also considerable interest in wireless VoIP technology by enterprises around the world.

Just like wireless voice, wireless data communication can also be categorized into two categories:

*   Mobile wireless (Cellular-based): GPRS, EDGE, WCDMA (all GSM), and CDMA2000 (CDM).
*   Fixed wireless (Network based): WLAN, Wi-Fi, Bluetooth, and WiMax.

8.1.   Technology Standards Evolution

The wireless data technologies have been evolving over the years (refer 8.1) starting from 2G to the future 4G. The following are the technology generations.

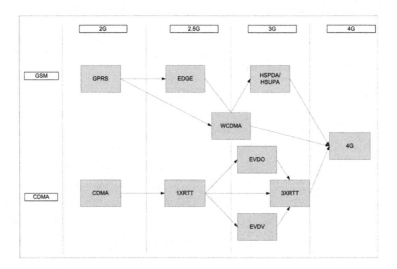

Figure 8.1 - Wireless Data Technology Evolution

8.1.1.   1G

The 1G network provided an analog voice service, but was not capable of providing data service.

8.1.2. 2G

2G was the first wireless technology standard that supported data communication but still it was quite limited. The top speed achieved by 2G technologies is up to 14.4 Kbps. However, such bandwidth speeds is woefully inadequate compared to the requirements to qualify as true broadband (in the order of 256 Kbps or higher for download). To overcome these deficiencies, both GSM and CDMA have developed enhanced technologies that can provide higher bandwidth.

8.1.3. 2.5G

2.5G is the most widely implemented technology around the world today (2008), and is essentially an intermediate standard between 2G and 3G phones. 2.5G technology involves a software enhancement over existing 2G technologies, making data communication packet-based. 2.5G also has a much higher bandwidth speed of up to 56 Kbps. A new wireless standard, Enhanced Data GSM Environment (EDGE), has been developed to increase the bandwidth to a peak rate of 384 Kbps, allowing GSM operators to offer high-speed data communication services.

8.1.4. 3G

The third generation (3G) of wireless technologies promises better voice quality, much higher data communication speeds, and a host of multimedia services. The ITU (International Telecommunication Union – a UN body) mandates that 3G networks, among other capabilities, deliver improved system capacity and spectrum efficiency over the 2G systems and support data services at minimum transmission rates of 144 kbps in mobile (outdoor) and 2 Mbps in fixed (indoor) environments.

Currently only CDMA's CDMA2000 and GSM's WCDMA qualify as 3G standard. Because 3G networks use different transmission frequencies and require different infrastructure from 2G and 2.5G networks, operators and carriers are incurring significant expenses (in the billions of US dollars worldwide) in order to buy spectrum licenses and develop their 3G infrastructure.

8.1.5. 4G

4G is the next generation of standards and is not yet fully developed. In the simplest terms, 4G is the next generation of wireless networks, which will replace 3G networks at some point in the future. 4G is expected to address the limitations and problems of 3G in terms of performance and throughput. For example, 3G is not capable of meeting the needs of future high-performance applications like multimedia, full-motion video, or wireless teleconferencing.

4G standards are likely to recommend an all-IP network and offer peak data communication rates of up to 100 Mbps while roaming and 1 Gbps for fixed. It plans to utilize IP in its fullest form with converged voice and data capability.

8.2. Mobile Wireless Technologies

Just like voice, there are two key families of mobile wireless data communication technologies:

- GSM-based, which includes GPRS, EDGE, WCDMA, and UMTS.

- CDMA-based, which includes CDMA 2000 1X, CDMA EVDO, CDMA EVDV, and 3XRTT.

8.3. GSM Technologies

8.3.1. GPRS

General packet radio system (GPRS) is a 2G GSM-based technology that is extensively used in GSM phones for data communication purposes. GSM is used to support voice communication while GPRS supports data communications, and they can both run on the same device. A data communication rate of up to 115 Kbps is achieved by using GPRS technology. The most important features of GPRS technology are as follows:

- Packet-switched, which means that the spectrum is only used when necessary
- Cheaper as users do not pay by the minute, but only pay for what they transmit or receive
- Connection is "always on"
- Migration to the next generation of data communication technology like EDGE, WCDMA and 3GSM is easier

8.3.2. EDGE

The need for speed is always growing, especially when it comes to broadband data communications. GSM's answer to increased data communication speed is "Enhance Data rate for Global Evolution" (EDGE) technology, which is considered 2.5G technology. GPRS packet data is an inherent part of most current mobile operations, but the implementation of EDGE to provide higher speeds of data communication is fast catching up among GSM operators. EDGE technology is an overlay on existing GPRS technology, which enables each base station to carry more voice and/or data traffic by using special coding schemes. The data communication speed achieved by EDGE-enabled networks is up to 384 Kbps. The most attractive feature of EDGE is that it is just a software upgrade to existing infrastructure at base stations, hence easy to implement and inexpensive for the operators. In addition, just like GPRS, EDGE helps in migrating existing infrastructure to 3G technology.

8.3.3. WCDMA

WCDMA stands for "Wideband Code Division Multiple Access". It was developed by NTT Do-CoMo, and it uses the generic code division multiplexing technique. It is accepted by ITU as 3G standard (IMT-2000) and is now part of UMTS (3G successor to GSM, discussed in 8.2.4 section below). It was designed to provide cost-efficient capacity for both modern mobile multimedia applications and mobile telephone services.

The advantages of WCDMA are faster data communication speeds of up to 2 Mbps (peak at stationary) and integrated support for voice communications. With this technology, users will be able to do multiple tasks, like speak and see the person live on video, or speak and surf the Web at the same time.

8.3.4. UMTS

Universal Mobile Telecommunication Systems (UMTS) is a set of voice and data standards based on GSM and WCDMA technology. Wideband CDMA (WCDMA) is the radio technology used in UMTS. As a result, the terms UMTS and WCDMA are often used interchangeably. UMTS is also referred to as 3GSM, emphasizing the combination of the 3G nature of the technology with the GSM standard, which it is designed to succeed.

Key features of UMTS are as follows:

- IP-based technology capable of supporting peak data rates of 350 Kbps when the user is mobile
- Designed to support bandwidth-intensive applications such as streaming multimedia, large file transfers, and videoconferencing
- Designed to support delay-sensitive applications such as virtual private networks (VPNs) and real-time multiplayer gaming
- High spectral efficiency for voice and data, support for simultaneous voice and data
- Low infrastructure costs
- Global roaming
- Broad selection of handsets and other user devices

UMTS is truly capable of making mobile broadband as good as fixed broadband. It will play an instrumental role in positioning 3G/UMTS as the preferred choice for broadband for both mobile and fixed usage.

8.3.5. HSDPA

HSPDA (high-speed downlink packet access) and HSUPA (high-speed uplink packet access) are enhancements of WCDMA and promise download speeds of up to of 14.4 Mbps and upload speeds of 5.8 Mbps, clearly much higher than the speeds promised by 3G standards. This will enable operators to provide advanced services at lower costs. These technologies are also supposed to be backward compatible to WCDMA and will not be very expensive to upgrade. The goal of providing such high speeds is to enable consumers to conduct data-intensive operations like e-commerce and support wireless VoIP to bring in more revenue to the operators.

HSDPA/HSUPA is considered to be a 3.5G technology.

8.4. CDMA Technologies

8.4.1. CDMA 2000

CDMA 2000 is a CDMA-based 3G mobile telecommunications standard that represents a family of technologies. Today, it is collectively managed by the 3GPP2 standards organization.

The CDMA 2000 family includes the following technologies:

- CDMA 2000 1XRTT – This is capable of doubling the voice capacity of CDMA networks and delivering peak packet data speeds of 307 kbps in mobile environments
- CDMA2000 1X EVDO/EVDV – The next step in the evolution of CDMA2000 1XRTT is labeled as CDMA2000 1XEVDO/EVDV. The 1XEV plan is being implemented in two phases:
  o Phase 1 is 1XEV-DO (Evolution – Data Only) supports a downlink peak data communi-

cation rate of up to 3.1 Mbps and uplink rates of up to 1.8 Mbps. 1XEV can support high bandwidth applications like MP3 music download, e-commerce, and video conferencing
   o   Phase 2 1XEV-DV (Evolution – Data and Voice) supports much higher peak data speeds of up to 5 Mbps
- CDMA 2000 3XRTT – CDMA 2000 3XRTT uses three channels of 1XRTT to provide service, which enables it to provide much higher speeds

8.5.   Fixed Wireless Technologies

Fixed wireless data communication technologies use the same network concept discussed in chapter 5 but without the hassle of wires.

8.5.1. Bluetooth

Bluetooth is a wireless technology mainly used in wireless personal area networks (PAN) for conducting device-to-device communication. In general, it is used to establish wireless communication between closely placed electronic devices. The data transmission rate in Bluetooth is only 300 Kbps and only has a range of around 32 ft. However, unlike the Wi-Fi technologies described above, Bluetooth requires no adapters, routers, gateways, access points or complex setup schemes in order to connect devices. Any Bluetooth-enabled device can communicate with other Bluetooth-enabled devices with little or no preparation.

8.5.2. Wi-Fi

Wi-Fi (wireless fidelity) is a wireless technology mainly used in wireless local area networks (LANs – refer chapter 5 for details). LANs allow computers located within a home, office building, or school campus to communicate and share resources (broadband connection, printers and databases) with each other. For example, a computer is usually connected to the broadband modem using wires, but Wi-Fi-enabled modems and wireless devices can eliminate wires, allowing free mobility within a certain range. Wi-Fi has caught on like wildfire – consumers are installing Wi-Fi enablers on top of their broadband modems and buying Wi-Fi-enabled computers. This move has enabled them to browse the Internet from anywhere in their homes or connect to their office network from airports, coffee shops, and other places. Wi-Fi is a boon for heavy travelers who want to access the office network or Internet from anywhere.

How It Works
The working concept of Wi-Fi is very similar to that of a cell phone, except that there are no cells. There is a base station called a "hotspot", which is nothing more than a wireless transmitter and receiver. Wi-Fi-enabled computers can send and receive data as long as they are within the range of the base station. The key difference between cell phone cell sites and hotspots is that the hotspots are not as big like cell sites. The hotspot is a small box containing a low-powered transmitter and receiver, with a typical range of only 750–1000 ft. Another key difference is the speed – Wi-Fi technology supports data communication speeds in the range of 10–54 Mbps, depending on the standard used. This is several times faster than even wired broadband modems, and is definitely much faster than the data communication speed supported by cell phone technology (GSM and CDMA).

Every wireless service provider is in a mad dash to install hotspots. Most coffee shops, airports and downtown areas are already covered with numerous hotspots installed by different service providers. Even at homes, a Wi-Fi enabling device is connected to the broadband modem, which allows all Wi-Fi-enabled devices to connect without wires. In this case, the Wi-Fi enabling device acts as the hotspot, with similar speeds of 10–54 Mbps, but the range is limited to only 125–200 ft.

Advantages

The greatest advantage of Wi-Fi is that it allows mobility within premises (home or office) to for broadband connectivity. In addition, there are several other advantages:

- No cabling required
- Wi-Fi-certified products from different vendors work seamlessly, e.g., the Wi-Fi-enabled computer might be from Dell while the modem might be from Cisco
- Wi-Fi is a global standard and Wi-Fi-certified products work anywhere in the world
- Wi-Fi has strong security mechanisms in place

Disadvantages

Wi-Fi has certain disadvantages like:

- Popular cordless phones and Wi-Fi work on the same frequency of 2.4 GHz, causing annoying interference
- Configuring security on Wi-Fi installation sometimes proves tricky

Technology Standards Evolution

Over the years, Wi-Fi technology has evolved through different standards, starting from 802.11b. Each standard has different features and capabilities:

- 802.11b – This was the first standard set by IEEE and is the choice of enterprises today. 802.11b is the least expensive amongst its peers and has the widest range, making it suitable for use in big office spaces. It operates in the 2.4 GHz radio spectrum and has a data transmission rate of up to 11 Mbps within a 30 m range. On the other hand, 802.11b operates in the same frequency range as cordless phones, causing annoying interference
- 802.11a – This was introduced to overcome many of the shortcomings of 802.11b. It operates at 5.15 GHz and has a much higher data transmission rate of up to 54 Mbps. The biggest disadvantage over 802.11b is its shorter range of only up to 15 m
- 802.11g – This is the latest standard, and although it operates at 2.4 GHz, it is capable of supporting speeds of 54 Mbps at a range of 30 m (100 ft). 802.11g holds considerable promise and is likely to be the most widely used standard in the future

Criterion	802.11b	802.11a	802.11g
Speed	11 Mbps	Up to 54 Mbps	Up to 54 Mbps
Data encryption	128 bit WEP	152 bit WEP 256 bit AES	128 bit WEP
Max range	Up to 30 ft	Up to 10 ft	Up to 30 ft

Compatible with	802.11g	–	802.11b
Potential users	Home networks	Large enterprises concerned with security	Larger networks, small enterprise

Table 8.1 - Wireless data technology comparison

8.5.3. WLAN

Wireless local area network (WLAN) is the name enterprises give to Wi-Fi-enabled local area networks (LANs). Since no wires are required, it is easier and cheaper to set up a WLAN as compared to wired LANs. The incremental expense of adding computers is practically zero and can be achieved very quickly as no wiring is required. In addition, WLAN allows employees to access the network even when they are not at their desk, for example in a conference room for a meeting or at the most popular location – the coffee shop in the lobby.

8.5.4. WiMax

WiMax (Worldwide Interoperability for Microwave Access) is a wireless technology used in a wireless metropolitan area network (MAN). It is capable of supporting a variety of applications like Internet access, VoIP, IPTV and others, all at the same time. Theoretically, WiMax is expected to cover a range of up to 30–40 miles with data communication rates of up to 70 Mbps, but today it is practically capable of supporting speeds of 10 Mbps in a 3 km densely populated area or a 10 km rural area. Speed and coverage are expected to improve as technology improves. WiMax uses frequencies in the bands of 2.5 GHz, 3.5 GHz and 5.8 GHz.

IEEE has developed the 802.16x series of standards that govern the requirements for WiMax. WiMax technology is the biggest threat to current broadband technologies like DSL, cable broadband, 3G and Wi-Fi, and is being trialed by many companies around the world. With WiMax, a service provider can provide broadband service to an entire city without installing any cables or wires. In addition, WiMax is expected to find a lot of use as a backhaul for hotspots and cellular networks, especially those located in far-flung areas. One of the key advantages of WiMax is that it is technology neutral, so both CDMA and GSM companies will be in a position to exploit it.

How It Works

A simple WiMax setup will have a central tower broadcasting wireless signals to small WiMax-enabled dishes located at customer premises. The coverage area of each WiMax tower is referred to as a "warm zone" compared to the Wi-Fi's "hotspot."

Advantages:

- Can cover complete geographic area without requiring any wires
- Network setup is faster and cheaper
- Supports both mobile and fixed users
- Open standard
- Completely IP-based
- Capable of supporting both line of sight and non line of sight environments
- Strong industry support

In short, the future of telecommunication is WiMax. Even though the technology is in its nascent stage still (2008), but it has the potential to completely change the landscape of the business.

8.6. Wireless Security

Security concerns are the nemesis hindering the speedy adoption of wireless data communication. There have been lingering doubts about the ability of wireless infrastructure to provide the same kind of security as wireline infrastructure. In a wired world, the walls of the home or office provide sufficient protection to the network, but in a wireless world, radio waves are accessible within a certain range outside the walls as well. This fact is not lost on hackers – by staying in close proximity to the buildings and using various sniffing techniques, they have been able to get easy access to networks whose security features are not properly configured.

To overcome security concerns, both individuals and enterprises must tackle wireless security from two angles:

- Better understanding of existing security technologies
- Design an adequate and sustainable security policy and strictly follow it

8.6.1. Security Tools

The first of the security tools specified by IEEE in the 802.11b Wi-Fi standard is called "Wired Equivalent Privacy" (WEP), which is used quite extensively today. It was designed especially for WLAN, but is applicable to home wireless infrastructure too. WEP works by encrypting the data transmitted. Encrypted data along with existing security mechanisms like password protection and dynamic secure IDs ensure a strong security setup.

WEP has been in place long enough for people to find ways around it. The 128-bit data encryption that comes with WEP was thought to be impregnable during the early days, but now there are tools available free on the Internet that are able to crack the encryption with considerable ease. Of course, WEP was never considered the lone security mechanism for WLAN, but the fact that it is easily penetrable has added to Wi-Fi's woes.

"Wi-Fi Protected Access" (WPA) is the latest tool in Wi-Fi security. The biggest drawback of WEP is its use of a static key for encryption and no built-in authentication mechanism. WPA fixes these flaws by using dynamic keys for data encryption and forcing a strong mutual authentication mechanism to be in place. WPA uses a different key for every frame transmitted, which is a big improvement over WEP, where the keys are not changed for weeks or even months. The mutual authentication in WPA makes sure that the user communicating is a pre-registered authorized user of the system. Another advantage of WPA is that it is very easy to upgrade existing Wi-Fi-compliant components to use WPA.

8.6.2. Security Policy

No matter how good a lock is, it does not offer any protection if the door is not locked properly or if the keys are easily accessible to the wrong people. Security in the communications world is no different, regardless of whether it is a LAN, WLAN or home network. Many surveys conducted in big cities across the world show that anywhere from one-third to one-half of installed enterprise WLANs are penetrable. This percentage is as high as 90% or more in home networks. A combination of tools and technology is insufficient; what is required is a strong security policy, a good understanding of how to use the existing security tools, and even stronger discipline in following up.

It is estimated that by 2008, almost all laptops and handheld devices like PDAs will be Wi-Fi-enabled, but few corporations have a strong security policy in place. Simple security policy like locking down wireless-enabled systems when not in use, enabling WEP or WPA security protocol, installing personal firewalls, using VPN, and quickly reporting lost or stolen devices to the security department are not that hard to come up with. Apart from having good security policies, it is equally important to conduct extensive testing to make sure that they work. It is also important to have a centralized security organization that must devise and implement security policies and be the central point for speedily resolving security-related issues. If security-related issues are not quickly resolved, people tend to shut down firewalls or remove encryption as the first step while trying to resolve the problem themselves.

8.7. Mobile TV

Mobile TV refers to video services delivered to mobiles phones over the 3G and 3.5G mobile networks. It includes broadcast and unicast services like live TV, time-shifted TV, on-demand TV, sports, news, and music. Many of the popular TV sitcoms, news, traffic and weather now have a "one-minute" condensed version for people to watch on their mobile phones.

Mobile TV is still in its nascent stages. Recent technological developments are making TV and video services truly mobile, some of them are:

- Portable handsets with relatively large, high-resolution LCD screens, powerful CPUs and long battery life now provide users full viewing enjoyment with freedom of movement
- TV's transformation from analog to digital has led to huge advancements in video compression/decompression that have reduced bandwidth requirements for acceptable quality video signals
- Availability of good bandwidth
- Technology standards like DVB-H, IPDC, DMB and MediaFlo are making it possible to realize mobile TV

Market research indicates that many consumers show an interest in watching their favorite programs on their mobile devices, especially while commuting or waiting for the bus. Regardless of how people consume the content, mobile TV promises to help the bottom line of distributors and multimedia content suppliers by permitting premium services to be offered. A trial in Oxford, conducted by O2, has found that nearly 80% of people would subscribe to a mobile TV service. A similar trial from BT has shown that people would pay up to £8 a month for such a service. O2 was surprised to find that people were using the mobile TV service at home. According to their trial results, 36% of people used the service mainly at home, compared to 23% at work or university and 28% while on the move. The O2 and BT trials used different technologies but drew similar conclusions about the appeal of mobile TV to consumers. In addition, mobile TV is expected to open doors to a significantly large number of applications like YouTube, Video share and other user generated multimedia content on the mobile channel that will also increase the usage.

8.7.1. Key Drivers

Mobile TV is a promising new service that both consumers and service providers are eagerly looking forward to. The following are some of the many factors driving this growth:

- Reach: close to 1/3 of worlds population has a cell phone
- Lifestyle: most people are not in front of a TV during a major portion of their waking hours. This makes watching streamed video on mobile handset a worthwhile proposition. In addition, many people will prefer to 'snack' on short video clips instead of watching lengthy TV shows, especially on news, weather, traffic, and soap operas
- Flexibility: mobile TV will allow the user to choose what they want to watch, when they want watch and where they want to watch
- Interactivity: Applications like games, video share etc will allow customer to interact with other people
- Technology: new technologies like DVB-H, DMB, and MediaFlo are capable of supporting good quality video streaming

8.7.2. Challenges

Mobile TV is still in early stages and as such has several challenges, some of which are described below:

- Viewing habits: due to early stages of trials and deployment, it is not clear what content customer want to watch, what offerings (pay, ad supported, hybrid) best meet their demand, and where the viable revenue streams are
- Content availability: Lot of content is available for traditional broadcast and video on demand viewing. However, not much of those are suitable for the small sized screen on mobile, and the length of the programs also needs to be considerably reduced
- Technology maturity: lot of the technologies required to support mobile TV are still evolving
- Handsets: The current crops of handsets are not suitable for watching TV. Handsets need more screen real estate, battery, memory, and processing power.
- Investment: Mobile TV is an investment intensive service that requires full 3G network and additional technologies capable of supporting unicast and multicast streaming
- Regulation: no clear regulatory framework capable of supporting mobile TV, especially around the needs of additional spectrum to support mobile TV is in place yet anywhere in the world.

8.7.3. Technologies

There are already multiple technologies emerging for use in enabling mobile TV. Some of the prominent ones are:

- DVB-H (Digital Video Broadcasting for Handhelds) - available in Europe, US, South Africa and Asia
- S-DMB (Satellite Digital Multimedia Broadcast) - South Korea, Japan
- STIMI (Satellite Terrestrial Interactive Multiservice Infrastructure) - China
- MediaFLO - launched in US, trialed in UK and Germany
- ISDB-T (Integrated Service Digital Broadcasting) - Japan
- T-DMB (Terrestrial Digital Multimedia Broadcast) - South Korea, Germany

- DAB-IP (Digital Audio Broadcast) – UK

However, the fight for supremacy is between DVB-H and MediaFLO. The players in this fight are again GSM and CDMA. GSM is promoting DVB-H as an open industry standard, while CDMA (Qualcomm) is promoting MediaFLO.

DVB-H

The Digital Video Broadcasting (DVB) Project is an industry-led consortium of over 260 broadcasters, manufacturers, network operators, software developers, regulatory bodies and others from more than 35 countries that are committed to designing global standards for the delivery of digital television and data services.

DVB-H is largely based on the successful DVB-T specification for digital terrestrial television (DVB-T), adding to it a number of features designed to take account of the limited battery life of small handheld devices, and the particular environments in which such receivers must operate.

DVB-H is an approved standard since November 2004 for handheld equipment by ETSI (European Telecommunications Institute) and is being widely adopted around the world. GSM has wholeheartedly supported DVB-H as its technology standard for mobile TV.

DBV-H is also working on using Internet Protocol (IP) technology for transmission purposes. This technology is called "IP Datacasting". In this case, all content is delivered in the form of IP data packets, which results in efficient and cost effective distribution of digital content to mass audiences. Mobile phone users can access music, Web pages, and games, as well as television and radio.

MediaFLO

Media FLO™ is a proprietary technology promoted by Qualcomm. It is a comprehensive, end-to-end solution designed specifically to address the inherent challenges of efficiently and cost-effectively distributing mass volumes of high-quality mobile multimedia to wireless subscribers. It addresses the usability, network capacity, and device constraints typical of video delivery to a mobile handset. Designed from the ground up with mobility in mind, it delivers a compelling user experience for the mobile masses.

DVB-H or MediaFLO – Which is better?

It is too early to say which is better. Technically, comparing the performance of the two systems, FLO can fit 20 channels while DVB-H can fit only nine. It also provides a much quicker channel change than DVB-H and supports longer watch time. Even though the discussion on these topics is not conclusive yet, but there is no doubt that MediaFLO is technically superior in many aspects. However, history has proved several times that being technically superior is not the only criterion for success. For example, CDMA is technically superior to GSM, but GSM has over 80% market share. This is because GSM is an open standard and as such is cheaper in all aspects, while CDMA is a proprietary technology that adds licensing costs to operations. It will be interesting to watch this battle for supremacy going forward.

8.8.   Wireless Data @ 2010

True broadband is the most anticipated technology in wireless communications. The current wireless broadband technology (2.5G) has not really taken off due to disappointing performance, while 3G is still expensive. If wireless broadband is to succeed, it needs to provide robust high-speed connectivity to any device, at any place, at an affordable price. The problem with wireless broadband today is that it is scattered, with different incompatible technologies provided by different users at exorbitant prices. In order for wireless data to displace wired broadband technologies (DSL and cable broadband), they need to be capable of supporting the voice, video and data needs of home and enterprise customers. By 2010, it is very likely that 3G, 4G and WiMax will provide connectivity to all devices at any place to any device, at a cost-effective price. Consumers will buy one wireless service that can support voice, video (mobile TV) and data services.

9. Basic Concepts of IP Communications

Internet protocol (IP) based communication is the last of the big seven revolutions in the communications world so far. For years, the communications world has treated voice, video and data differently. There are different standards, networks, protocols and systems in place to provide each of these services. Voice is processed using analog technology and transmitted on a low-bandwidth network made of simple copper wire. Data and video are processed using digital technology and transmitted on a high-bandwidth digital network made up of copper wire and coaxial or optical fiber. Different protocols are used for each of these services and the supporting IT systems are different as well. The reason for treating each of these services differently is that they each came into existence at different times. As each new service came into existence, new networks, protocols and systems were developed to support them.

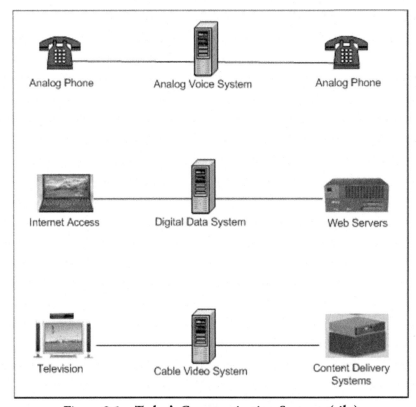

Figure 9.1 – Today's Communication Systems (silo)

However, the problem now is that maintaining and operating multiple networks and support systems is becoming difficult and expensive. In addition, since each uses different protocols, new convergent services (see chapter 16) based on these services are impossible. The answer to this problem is Internet Protocol (IP) based communications. IP does not treat voice, video and data differently, hence a single network, a single protocol, and a single IT support system can support all three services!.

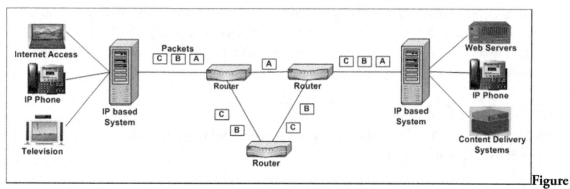

Figure

9.2 - IP Based Communication System

9.1.   What Is IP?

Internet protocol (IP) is a set of technical rules that define how two nodes (e.g., computers, routers) communicate over a network. There are currently two versions: IP version 4 (IPv4) and IP version 6 (IPv6).

IP mainly shot into prominence because of the Internet. Transmitting data using IP is nothing new. Consumers have been able to do that since the 1970s – but using IP for voice and video is only very recent. Voice was first transmitted using computers by a small communication company called VocalTec from Israel in 1995, and IPTV (television over the IP network) was made commercially available in the same year.

The base protocol in IP-based communication is TCP/IP (transport control protocol/Internet protocol), initially developed as part of the ARPANET project funded by the US Department of Defense. Work on the protocol began in 1973, and the first formal standard version of TCP/IP, version 4, used extensively in networks today, came out in 1980.

The market for IP-based communication is now set for growth. Technology is ripe, many governments are addressing the required regulations, demand is rising, and suppliers are jumping in. By the end of 2007 there were 70 million VoIP (voice over IP) subscribers around the world, with the annual growth rate at a phenomenal 79.8%. By 2010, the number of subscribers is expected to increase to 151.2 million. The number of households around the world subscribing to IPTV services offered by telecommunication carriers will reach 50 million by 2010.

9.2.   How Does IP Communication Work?

In traditional communication systems, every type of signal (voice, video and data) is processed, handled and transmitted in a different way. The switching techniques are different, the protocols used are different, and the way signals are transmitted is different for each type of service. This is because each one was developed in a different era, and each has evolved within its own infrastructure. However, in IP, regardless of the source signal, the information is digitized and bundled into equal-sized packets with enough intelligence built into them to reach their destination without requiring an expensive dedicated circuit between the sender and the receiver. The complex logic built into the IP network routes the packet through the best and fastest path to the destination. At the receiving end, the packets are reassembled in the correct order and delivered to the destination terminal. IP networks are also capable of prioritizing packets based on applications. This prioritization ensures that real-time traffic like voice, video is adequately supported, and quality is maintained. On the other hand, in a traditional circuit-switched network, an expensive dedicated path is established between the sender and the receiver.

9.3. Benefits of IP

The popularity of IP is because of the enormous benefits it brings to the table. Following is a list of some of the important benefits:

- Low operational costs: service providers and users see enormous cost benefits IP utilizes communication channel far more efficiently than traditional ways
- Low capital expenditure costs: single protocol is capable of supporting all three major groups (voice, video and data) of services. Thus, one system, one infrastructure, one set of people can support all three services
- Convergence: since a single protocol can support all services, it is possible to created converged services (e.g., caller id on your TV)
- Intelligent network: IP based network are very intelligent. During regular obstacles like congestion and network failure, the traffic gets rerouted automatically so that the packets still reach their destination. In addition, the advanced diagnostic capabilities makes it easy to quickly understand and fix problems before any major disruption to service occurs
- Scalability: it is very easy to scale an IP network without causing any interruption to existing services
- Channel agnostic: IP technology can support services over wireline, wireless, cable and satellite channels
- Security: advanced protocols like IP-Sec provide a very high level of encryption for data transmission

9.4. IP Services

The three key services provided by IP-based service providers are as follows:

- Voice over IP or VoIP (voice)
- IPTV (video)
- Broadband (data)

9.5. VoIP

Voice over IP or VoIP is defined as the transmission of voice over the Internet. It converts analog voice signals into digital packets and sends them over the Internet to a receiver at the other end. The receiving end can be an IP phone, a traditional phone over a PSTN network, or even a wireless phone.

VoIP users can be divided into three categories:

- Residential
- Enterprises
- Carriers

Consumers who have the Internet at home have been conducting some form of voice communication over the Internet using PCs for a long time, but VoIP allows them to communicate using a traditional phone or IP phones. In addition, the quality of service while using VoIP is far better than the service obtained by using PCs over the Internet.

Enterprises have been aggressively jumping on the VoIP bandwagon by converting their expensive PBX (private branch exchange) services to VoIP-based services. These services offer very high quality and provide many additional rich features that were unheard of before.

Carriers have been directing a lot of their long-distance back haul traffic over the IP network for a while now. It is distinctly possible that a consumer might have a regular PSTN connection but the carrier may be directing the traffic over the Internet.

9.5.1. How Does VoIP Work?

VoIP is a technology that digitizes voice signals, bundles them into packets, and transports them over a public or private IP data network. Voice traffic can originate and terminate from IP-based devices like an IP phone, a computer with a microphone and voice card, or an analog phone with a terminal adapter, from anywhere to anywhere in the world. The packets travel to their destination taking different paths over the Internet. At the receiving end, the packets are reassembled in an orderly manner.

A simple VoIP network (Fig 9.3) has four key components:

- Gateways
- Call controllers
- Media servers
- IP network

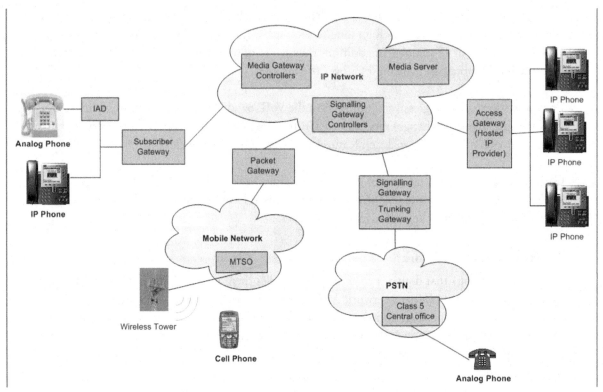

Figure 9.3 – How Does VoIP Work?

Devices (analog phone, IP phone, and computers) are connected to the IP network using media gateways. The IP network has call controllers that route the call to the right destination device using the right signaling protocol. Once a call is established, the call controllers drop out of the picture and the packets flow from source to destination using the fastest and shortest paths. If the destination is a mobile phone or an analog PSTN phone then additional gateways are used to establish connection.

9.6. VoIP Components

9.6.1. Gateway

A gateway is a device in the network that connects two networks working on different protocols. Without the gateways, signals cannot flow through from one type of network to another and as such, no communication between different networks is possible. For example, a gateway is required for devices on IP Network like VoIP phone to connect with phones on PSTN or mobile network. Gateways interconnect these different networks so that end-user devices can talk to each other. They convert and process signals as required and perform many other auxiliary functions like compression, echo cancellation, and statistics gathering. All gateways discussed below, except subscriber gateway, are located in the service provider's network:

- Subscriber gateway – Provides old analog and IP telephones with an interface to the VoIP network
- Access gateway – Provides a VoIP interface to analog/digital PBX
- Trunking gateway – Connects the IP network to PSTN

- Signaling gateway – Signaling is the process of generating and exchanging information between devices to establish, maintain and terminate connections between them. IP networks use C7 signaling protocol, while PSTN uses SS7 signaling protocol. A signaling gateway is used to convert the signaling used in the IP network to the signaling used in PSTN. Signaling and Trunking gateways are used together to connect the IP network to PSTN

9.6.2. Call Agents/Controllers

Call agents or controllers are responsible for establishing, maintaining, terminating and forwarding calls. They:

- Translate the phone number
- Look up the destination host
- Provide translation between different signaling protocols
- Maintain the session of the call
- Manage the resources that are being used during the call
- Additional services like caller ID and call waiting, and interact with other applications to provide services that are not hosted by them
- Support billing processes by providing usage information

9.6.3. Media Server

Media servers provide many value-driven advanced features to users like conferencing capabilities, IVR processing, broadcast messaging, text-to-speech conversion, speech recognition, language translation, and many more.

9.7. Basic VoIP Terminology

9.7.1. Enum

Devices connected to the Internet have unique identifiers called IP names. However, telephones connected to an IP network for VoIP service are also treated like any other device on the Internet. This means that each phone will have an IP address like 204.453.8.45, but this cannot be used as a telephone number. To overcome this, Electronic NUmber Mapping (ENUM) directories are created. In this directory, all VoIP-based telephone numbers are mapped to the IP addresses of the devices. Thus, when you are dialing to a VoIP phone, the right destination is identified by mapping the telephone number to the IP address from the ENUM directory. The greatest advantage ENUM mechanism provides is that a user can take his phone any where in the world where broadband is available and still be reachable on the same number mapped to its IP address in the ENUM directory.

9.7.2. IPv4

IPv4 is the fourth in the series of the IP protocols, and currently the most widely used version. One of the main goals of IPv4 is to provide unique global addresses to all devices on the network so that they can be identified. IPv4 uses 32-bit (4-byte) to create unique addresses, which results in over 4 billion possible unique addresses. This sounds like a large number, but remember that every machine on a TCP/IP network (a local network or the Internet) around the world needs a unique

IP address. Thus, it is inevitable that someday this pool of addresses will be exhausted. The next version of IPv4 called IPv6 overcomes this limitation.

9.7.3. IPv6

IPv6 is the successor protocol for IPv4. One of the main limitations of IPv4 was the limited number (4 billion) of unique addresses available for use to the world. IPv6 overcomes this limitation by using 128-bit addressing rather than IPv4's 32-bit addressing. This provides an almost unlimited number of unique addresses (2^{128}).

In addition to increasing the number of unique addresses, IPv6 also addresses some of the drawbacks of IPv4:

- Better supports for multicast transmission, which is very badly needed by IPTV operators
- Better QoS support that is badly needed for real-time applications like voice and video applications
- Native information security framework for both data and control packets
- Enhanced mobility with fast handover, better route optimization and hierarchical mobility

Few in the industry would argue with the principle that IPv6 represents a major leap forward for the Internet. However, it will be a major challenge to have an IPv6 address for all devices as currently almost all devices around the world use IPv4. It will be years before complete migration from IPv4 to IPv6 happens.

9.8. VoIP Signaling Protocols

Signaling protocol is the language spoken between two communicating devices. The two prominent protocols used in VoIP today are:

- H.323
- Session initiation protocol (SIP)

9.8.1. H.323

H.323 is the first of the VoIP protocols, and is approved by ITU as the standard for voice, video and data traffic over the IP network. Apart from the early start, the advantage of H.323 is the flexibility it provides for multimedia communications over the Internet and the easy integration with PSTN. It was also the first popular protocol that addressed the key issue of delay-sensitive communication by ensuring higher priority for real-time communication services like voice and video over the Internet. The biggest disadvantage is that it is considerably heavier as it uses a lot of signaling, even for a basic call in a simple network. As the networks become complicated with multiple gateways and controllers, and applications performing complicated operations like conferencing between different devices, the service quality running on the H.323 suite of protocols degrades considerably. The general trend in the industry is to migrate all existing H.323 establishment to SIP protocol.

9.8.2. Session Initiation Protocol (SIP)

Session Initiation Protocol (SIP) is a protocol that initiates and manages interactive user sessions involving voice, video, data sessions. It is a IETF approved 3GPP (Third Generation Partnership Project) signaling protocol. It is one of the major signaling protocols used in Voice over IP (VoIP). The advantage of SIP is that it is faster as it has a smaller code base, and uses a lot less signaling. These features make SIP-based systems simpler and more cost efficient than H.323-based systems.

The key difference between SIP and H.323 is the approach; H.323 is a one-stop shop for IP communications, with fixed solutions for handling all the problems, while SIP has very limited functionality by itself. The basic functionality provided by SIP is to:

- Locate the receiver
- Set up the session
- Provide call and feature management services
- Tear up the session at the end

The rest of the operations like resource reservation, quality of service, and security are provided by other industry standard protocols. The advantage of this modular approach is that one of the protocols used can be easily replaced if a new, better one is found in the market. The popularity of SIP has attracted lot of "add-ons" making it possible to use SIP for applications beyond voice, like data and video.

Some of the features that have made SIP so popular are as follows:

- Peer-to-peer protocol, requiring no implementation in the network level
- Independent underlying network protocol making it easy to ride over a variety of protocols
- Users can carry their SIP-based phone anywhere they go
- Ability to connect to users using email-style addresses
- Easily supports additional service features like call forwarding for voice
- Easy integration with the Internet, thereby making telephony another application on the Internet
- SIP is scalable, easy to implement, and requires less setup time as compared to other protocols

9.9. Advantages of VoIP

Apart from wireless voice, VoIP is the fastest growing voice communication mode. The following are some of the advantages of VoIP over other forms of voice communication:

- Reduced cost
- Convergence
- Mobility, can take VoIP phone anywhere in the world where broadband connectivity is available
- Rich features like find me – follow me, click to call, soft phone, area code selection, etc

9.9.1. Reduced Cost

VoIP technology has benefits for all users. Residential users can cut down communication costs by eliminating their telephone line and paying very low rates for national and international calls using VoIP phones. All calls are routed through the Internet so it doesn't matter where the destination is.

9.9.2. Convergence

One of the biggest advantages of VoIP is device convergence. Voice is treated as data in VoIP, which makes it easier to ride the same voice communication on any device, like a computer at one location, video conference equipment at another location, or IP phone or wireless devices at other locations. Emails can be read over the phone, or a user can be called from an address book stored on a computer.

9.9.3. Mobility

The mobility features provided in VoIP-based communications are impossible to provide using traditional analog phones. Users can carry their phone equipment anywhere without the caller ever knowing where the receiver is. A VoIP phone bought in the US or UK easily works in India or China. In addition, users can set up "find me – follow me" roaming services, whereby a call tracks down a user using a preconfigured set of phone numbers.

9.9.4. Rich Features

One of the key things that VoIP provides is rich features. Many features like call waiting, call forwarding, voice mail, call transfer, caller ID, call block, and so on are all free, and the best part is that they can be managed by the consumer without ever having to be dependent upon the service provider. Users can dial from a computer screen or get a screen pop-up to see who is calling. An un-limited number of auto-attendants, one for each group of caller, can be set up at no additional cost.

9.10. Disadvantages of VoIP

Even though VoIP seems very attractive, it still has the following issues to be addressed before it can fully replace PSTN:

9.10.1. Latency

Latency is the time it takes for voice signals to reach their destination. In circuit-switched com-munication, latency is negligible as a dedicated circuit is set up between the caller and receiver, but VoIP has some latency built into it due to the very nature of the setup. The caller's voice is chopped into packets and sent over the network, and the packets traverse different paths before being reas-sembled at the receiving end, which may take time. Latency depends on the physical distance between the users, the number of router hops, encryption, and voice/data conversion activity. A latency of around 80–150 ms is considered acceptable. Newer protocols have reduced the latency considerably, but it is yet to reach the level of quality provided by PSTN, which provide very low latency rates.

9.10.2. Jitter

Jitter is a variation of latency, and is measure of time between when a packet is expected to arrive and when it actually arrives. Packets are expected to arrive every 20–30 ms, but some packets do not for a variety of reasons, causing irregularities in communication.

9.10.3. Packet Loss

It is distinctly possible that many packets will get lost during transmission. In general, a packet loss of 1% is considered acceptable, but anything beyond 3% degrades the voice quality considerably.

Packet loss is an acceptable phenomenon in data communication where the receiving system is capable of requesting lost packets from the sender once again, but in time-sensitive communication like voice, there is no time to do so.

9.10.4. Reliability

Reliability is an important issue when considering VoIP. Traditional analog phones have attained close to 99.999% availability. Even though VoIP networks are often reliable, the fact that many different types of systems and technologies are involved creates additional points of failure.

9.10.5. Security

The Internet is notorious when it comes to security matters. Hackers have run amok and brought down entire networks by introducing simple malware (worms, trojans and viruses) into business systems that are accessible through the Internet. In addition, adwares and spywares are capable of invading users' privacy. All these issues will affect VoIP in varying degrees, but in general, voice communication over the Internet is far safer than data communication, even though security is not yet foolproof.

9.11. IPTV

TV has generally been considered to be the stronghold of terrestrial broadcast, cable and satellite companies, who control almost all the market. However, since 2005, leveraging advances in technology, traditional wireline telecommunication companies that offer broadband are aiming to enter the TV market by offering IPTV services. This is in direct response to cable operators who are offering voice services. Many telecommunication companies feel they have no choice - they must offer TV services to survive.

IPTV is very different from the standard broadcast/cable/satellite TV system, where you have to watch programs when they are being telecast. In addition, there is no interactivity between the user and the service provider; it is just a dumb one-way service. There is a clear indication from consumers that they want a TV experience that is more flexible than current options. CSMG ADVENTIS, a research firm indicates that consumers are dissatisfied with their current TV experience: more than 50% of consumers miss more than one program per week that they want to watch, while 85% frequently find there is nothing on TV they want to watch. However, with IPTV the viewer has complete freedom. Features like live and time-shifted TV, personal video recording, and video on demand give customers total flexibility to choose what and when to watch. In addition, graphical applications and interactive program guides let users interact with a service. Online tools like photo/video sharing allow them to share memories with family and friends, and network gaming lets a user play games with other gaming enthusiasts. IPTV also allows communication tools like telephone, instant messenger, emails, voice mail, and address book to run on TV. For example, you can store you phone book on TV and dial out from there. When you receive a call on your home phone, you can get the caller id on your TV and also the option to either pick the phone or let it go to voice mail.

IPTV is attracting huge investment from traditional telecommunication companies like ATT, Swisscom, Telecom Italia, Verizon, Telstra and BT. Without a TV service, traditional telecommunication companies are at a significant disadvantage as cable companies are able to offer a complete triple play (voice, video and data). For years, there was no choice for telecommunication companies, but recent advancements in DSL and FTTH technology, coupled with progress in video compression like MPEG4, are making IPTV a reality. With so many significant advantages from IPTV, cable companies are also keenly looking at adapting this technology.

By late 2007, there were close to 8 million IPTV subscribers around the world, with Western Europe being the main market. It is expected that by 2010, this will be a $10 billion market with over 40 million customers around the world.

Finally, IPTV is not to be confused with watching television through the Internet. Google, Apple, Yahoo! and others have begun to offer a service whereby a consumer can use a PC to watch a show that is mainly distributed through the open Internet. IPTV is run on a closed network belonging to the service provider. At no point during transmission do the packets use the Internet to reach their destination. This allows the service provider to offer a far better viewing experience, control the channels offered, and generate revenue as well.

9.11.1. Benefits of IPTV

The benefits of IPTV compared to traditional TV are as follows:

- On demand rather than broadcast
- Unlimited and varied content (live and time-shifted TV, video, music, games, etc.)
- Better picture quality
- Highly interactive
- Multi-language support
- Convenient (find, get, record, watch TV/videos when and where you want)
- Targeted advertising/merchandising
- Converged communication – IPTV can support communication services: phone, instant messenger, chat and multimedia, photo and video share
- Support for long-tail content and Web 2.0 applications

The biggest attraction of IPTV is the interactive experience that users can enjoy. Traditional TV transmission is one-way communication, whereby users can just watch the programs that are being broadcast at that time, but IPTV is two-way communication. Users can browse the program guide and select what they want to watch. They can participate in the show they are watching by voting, or buy a pizza by a click on the remote. Tele shopping will be as simple as shopping online. Users can play games with their friends or chat with them while watching a show. In addition, because IPTV uses the same kind of technology as VoIP and broadband, it will also be capable of converging different telecommunication services. Users can get caller ID, place a telephone call from a contact list, and video conference, all through the TV.

9.12. IPTV Services

A thorough understanding of IPTV is only possible via an understanding of the plethora of services available. Unlike cable, satellite or broadcast TV, which have few fixed services, IPTV makes previously unimaginable services possible. Some of these exciting services are listed below.

9.12.1. Broadcast TV

Broadcast service is an essentially continuous stream of video flowing from the content provider to the entire customer base, and is essentially the same as what you see in cable, satellite and broadcast TV. This includes channels like CNN, BBC, Discovery, etc.

9.12.2. Digital Terrestrial Television (DTT)

DTT is about transmitting broadcast TV signals using digital technology. These signals are received through a conventional antenna (or aerial) instead of a satellite dish or cable connection. DTT signals are transmitted by a nearby VHF or UHF transmitter and received via a set-top box that decodes the signal received using a standard aerial. DTT provides a clearer picture and superior sound quality when compared to analog TV, with less interference. In addition, DTT offers far more channels, providing the viewer with a greater variety of programs.

The UK has the highest number of DTT subscribers in the world, with DTT used for the popular set of channels that are collectively referred to as "freeview." DTT is extensively used across Europe and is very popular. Most IPTV providers in Europe also support DTT service over their set-top boxes.

9.12.3. Video on Demand (VoD)

Consumers spend billions of dollars on renting content like movies, and probably millions more in late fees. Renting movies is expensive and lot of work. VoD is a mechanism that allows customers to order a movie with just a few clicks on the remote, and then the movie or any other content like a boxing match will be rendered directly over the TV set. The service is provided by storing content on a server and streaming it to the viewer at their request. The user has the option to stop, start, fast forward or rewind the video, just as if they were watching a movie on their DVD player. Users can also watch the content as many times as they want over a fixed period or even buy it.

This is an extremely attractive proposition for renting movies as the customer does not have to go anywhere to rent a movie, nor do they have to worry about late return penalties. VoD is also an extremely attractive revenue opportunity for IPTV operators because of the strong demand among existing residential subscribers. In addition, VoD is useful in many other applications like sports, music, games, e-learning, training, marketing, entertainment, and other areas where the user would like to view programs at their convenience. It is a key revenue-generating stream for service providers and has the potential to increase the average revenue per user (ARPU) for the existing residential customer base by as much as 25%.

9.12.4. PPV

Pay per view (PPV) is an offering of paid television programs whereby customers can buy a particular program separately from any package or subscription.

9.12.5. Personal Video Recorder (PVR)

IPTV set-top boxes come with an in-built PVR. A PVR is a device that is capable of recording a video stream so that the user can watch that show at any other time, or even watch one show and record another. It is capable of scheduling a TV recording, viewing and pausing live TV, and acting as a media center to watch movies, listen to music, view pictures, listen to FM radio, and record multiple shows at the same time. Intelligent series linking will allow program episodes to be recorded based on preferences, first-runs, repeats, or all occurrences. They also allow remote scheduling of recording through the Internet or a mobile phone.

9.12.6. Enhanced TV

IPTV makes it possible to view many channels on a TV at once. This way, a sports fan will be able to keep an eye on six games at once on the same screen. In addition, a user can see a game from any of the cameras on the ground, not just the one that the provider is broadcasting.

9.12.7. Interactive Applications

Interactivity is the most important concept of IPTV, a feature that current cable, satellite and broadcast providers lack. Unlike cable, satellite and broadcast TV, IPTV is a two-way communication channel. Users don't have to just sit and watch – they can participate as well. This enriches the end-user experience and opens up a wide spectrum of possibilities. It enables exciting applications like t-commerce, search and recommendation, voting, communication applications like instant messaging, chat, interactive advertising, real-time betting, and single and multi-user network gaming.

9.12.8. Personal Content

There are quite a few interesting applications for customers to share personal content through IPTV. For example, people can securely share personal photographs and videos with their family and friends.

9.12.9. Social Applications

One of the most important developments in the world of IPTV is the support for social applications. It includes applications like dating, personal profile, little league, local league, social networking (like Facebook/Orkut but customized for TV), message boards, ratings and previews of movies and restaurants, and many more.

9.13. How does IPTV work?

First things first, IPTV does not mean that people will log into a Web page and watch TV, nor is it a reality show, as some people thought when asked in a recently conducted survey! Yes, IPTV does use the same protocol as that used in accessing the Internet (IP), but it doesn't use the Internet at any point for transmitting video packets from source to destination. IPTV service providers provide a TV service over a closed network with no connection to the outside world from the consumer end.

As shown in Fig. 9.4, video streams like live TV, time-shifted TV, video on demand content (movies and matches), interactive content like games, shopping and promotions, are bought by the service provider from many different content providers. All the streams are sent into one super headend where they are edited and any custom graphics is added. Once the content is ready, it is compressed and encoded into an MPEG4 stream of various bitrates, as required (SDTV, HDTV format). The video stream is then sent to multiple regional headends.

The regional headend is the workhorse of the IPTV service, where the bulk of the IPTV middle-ware platform is installed. It is responsible for the following activities:

- Service management
- Subscriber management
- Catalog management
- Order management
- Billing management
- Electronic programming guide
- Video delivery server
- VoD server
- Encryption
- Ad insertion
- Scheduler
- Encoder

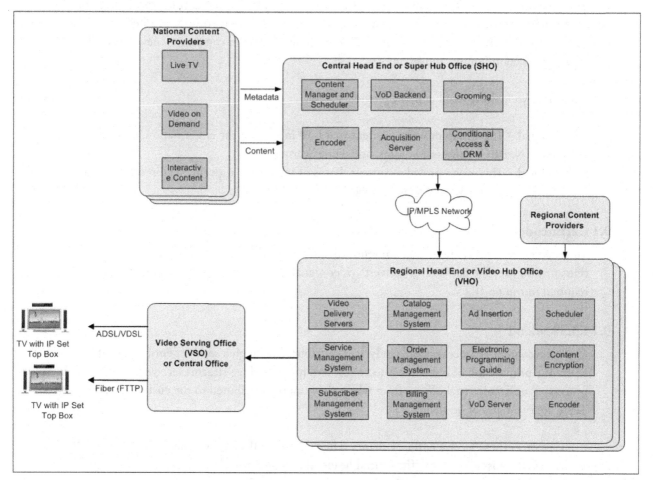

Figure 9.4 – How Does IPTV Work?

The key here to note is that the entire network from central headend to the local central offices is a closed network owned and managed by the service provider. No part of transmission goes over the Internet at any point. This allows the service providers to manage the entire transmission so that a customer can see TV of the same quality as the competing cable, satellite and over-the-air broadcasting services. If the Internet is used then the service provider has no control over the traffic flow, which degrades the user experience due to latency and loss of packets. Since it is a closed network, the provider is capable of prioritizing the video traffic throughout the network so that acceptable quality of service (QoS) is maintained and there is little delay and loss of packets. Unlike the Internet, a closed network also ensures that only authorized users are allowed to view copyrighted content, and revenue can be generated by making the services payable.

9.14. IPTV Components

This section describes the various players and components involved in the delivery of IPTV services (see Fig. 9.4).

9.14.1. Content Providers

Content providers are the source for the content shown on TV. Content providers can be classified as providers of live TV, stored content (movies, time-shift TV), EPG (Electronic programming guide), educational programs, and interactive content (games, competitions, promotions). Within each category, there are various players like independent television companies, news networks, movie production houses, and so on.

There are two main types of content providers:

- National – This group provides content that is of interest to the entire nation. Good examples are CNN, BBC, Universal Studios, and Time Warner
- Regional – This group provides content that is of interest to a specific region only. Good examples are local news and sports providers

9.14.2. Encoder

The encoder receives content from content providers and converts it into a format suitable for transmission over the network. It converts video signals from various formats to MPEG4, which is required to run on IPTV.

9.14.3. Headend

Once the content is processed and ready, it is distributed through a series of entities called headends. The primary job of headends is to collect and forward video streams received from content processing units to central offices, from where they can be distributed to the customers' premises. There are two types of headends:

- Central – These receive national channels like CNN or BBC, movies and other content that is shared across all regions. The central headend is sometimes also referred to as the super hub office (SHO)
- Regional – These are set up per region (state-level). Regional and local headends are set up to receive regional/local programs. The regional headend is sometimes also referred to as the video hub office (VHO).

Depending on the size of the carrier and the geographic spread, there is usually one central headend and multiple regional headends.

9.14.4. IPTV Middleware Platform

The IPTV middleware platform is the heart of IPTV operations. It is a set of software components that helps in delivering content to the end user and controls the user interaction with the service. For example, the user interface, channel change, interactive actions like gaming, voting and services available to a consumer (such as the electronic program guide, VoD, or pay per view service), are all made available and controlled through the middleware. In addition, it manages content distribution, rights control, content security, and billing.

The following are the key components of an IPTV middleware system:

- Subscriber management system – The subscriber management system is located at the regional headend and stores customer information (profile) that includes name, address, and subscription details. This is the commercial heart of the system and manages access control, billing, and customer relationship management
- Service management system – The service management system allows carriers to define, configure and maintain services. All business rules regarding services like conditional access details (which customer should get which service), availability details (when and for how long), and cost (for how much) are stored in these systems
- Video delivery server – The main function of a video delivery server is to stream video to a user upon their request. It allows the customer to manipulate the rate at which a video is played (start, stop, pause, rewind and forward). Service providers usually have multiple servers configured in a server farm to deliver the maximum number of required streams and provide fault tolerance. Each server is capable of storing a large number of video files.
- CAS and DRM system – The digital rights management (DRM) system defines the digital rights of the content. It defines who can watch what content, when, and for how long by generating appropriate entitlement keys. These keys are used to decrypt the content stream and make it available for viewing via the set-top box. The conditional access system (CAS) implements the restrictions mentioned in the DRM system.
- Billing management system – This system is responsible for recording usage, which is sent to the billing system for invoicing.
- Electronic programming guide – This component acquires service listings data from various sources and prepares a program guide that allows viewers to see which programs are available.

Some of the major providers of IPTV middleware platforms are Microsoft, Kasenna (now Espial) and Myrio (Nokia-Siemens).

9.14.5. IP Set-top Boxes (STB)

An IP set-top box (STB) serves as an interface between a TV, the broadband network, and the video delivery system. It receives the MPEG stream from the IPTV middleware and converts it into a format compatible with the TV. In addition, they provide personal video recording (PVR), a local access control mechanism (parental locks), many interactive features, and certain gaming functionality as well.

Some of the major providers of IPTV set-top boxes are Motorola, Siemens, Phillips and Amino.

9.15. Basic Terminology

9.15.1. Digital Rights Management (DRM)

DRM is a mechanism used to protect the rights of content producers from unauthorized use of their copyrighted content material like music, films and text. DRM represents a family of technologies that defines access to copyrighted material. It is implemented through a conditional access system (CAS). DRM and CAS help content producers and distributors to prevent users from unauthorized use, illegal copying and distributing free amongst friends and family or on peer-to-peer sites like Kaaza and Napster.

9.15.2. Conditional Access System (CAS)

One of the major issues that content providers face is how to offer content selectively to subscribers. CAS is a solution that allows operators to set paid channels, pay per views, support password access control, and anti-copy control. It ensures that only those users who are authorized can watch the program.

CAS uses scrambling and encryption techniques. The video and audio streams at the operator's end are scrambled and a key is transmitted to the receiving set-top box. The set-top box is capable of descrambling using the keys in order to make the video and audio available.

9.15.3. Quality of Service (QoS)

Customers expect a very high level of quality from a TV service. Any delays, jitters, hazy picture will not be tolerated. Quality of service (QoS) is the mechanism used by service providers to maintain a very high quality of viewing experience. It involves very careful engineering of various parameters of transmission like bandwidth to ensure viewers get good quality signals.

9.15.4. Electronic Program Guide (EPG)

An EPG is an application that displays a list of current and scheduled programs on each TV channel for 14 days or so. It allows a viewer to navigate, select and discover content based on time, title, channel and genre by using a simple remote control. Many service providers are now using EPG to display the programs on Web sites and mobile phones.

9.15.5. Bitrate

Bitrate measures the average number of bits that one second of video or audio data will consume. Higher bitrate means bigger file size, but better quality. In a true sense, bitrate defines the amount of detail stored for video or audio per second.

Bitrate is a very important factor that providers have to deal with. Higher bitrate files mean better quality, but also mean that more bandwidth is required between the serving node and the customer premises.

Typical acceptable bitrates are as follows:
- Minimum recognizable speech is 4 Kbps
- Telephone quality audio is 8 Kbps

- CD quality audio is 240 Kbps
- Standard definition TV stream on MPEG2 is 1.8 Mbps, but around 0.9 Mbps on MPEG4
- High-definition TV stream on MPEG2 is 10 Mbps, and around 5 Mbps for MPEG4
- Uncompressed standard definition TV stream is 270 Mbps.

9.15.6. Compression

The bitrate of a typical standard definition video signal is around 200 Mbps. A typical DSL line has a maximum bandwidth of 8 Mbps. Thus, it is not possible to transmit a 200 Mbps raw video stream over DSL line. The solution is to reduce the number of bits sent down the channel without excessively reducing the quality. For example, since successive frames in a movie rarely change much from one to the next, so the MPEG2 compression technique only transmits the bits that have changed from the previous frame and retains the rest. This is a powerful technique capable of reducing the bandwidth required from 100 Mbps to around 3 Mbps for an acceptable quality of viewing. MPEG4 further reduces it by half to 1.5 Mbps for similar viewing quality. Thus, compression makes it possible to transmit video and audio signals despite bandwidth constraints.

9.15.7. Encoding

A digital video can exists in various formats like MPEG, AVI, QuickTime, Real Media, and Windows Media. DVDs follow the MPEG2 standard and VCDs follow the MPEG1 standard. Encoding is the process of converting a video and audio stream from one format to another so that it can be transmitted, stored and edited.

9.15.8. Codec

A codec or coder-decoder is a device used for encoding and decoding a digital data stream or signal. Incoming signals are usually coded in a different format so that they are suitable for transmission, storage or encryption. The same signal needs to be decoded for viewing or editing.

9.15.9. Metadata

Metadata is defined as "data about data", and is used to describe any type of data like videos, text and images. It may simply contain the title of the movie, the actors in the movie, a complete index of the different scenes in a movie, or provide business rules detailing how the content package may be displayed, copied or sold. The program details seen through the electronic program guide (EPG) on a TV are sourced from the metadata of the content stored.

9.16. Video Standards Technologies

The technologies involved in the compression and decompression of digital TV are evolving very rapidly. As mentioned earlier, a raw video signal has to be compressed quite heavily before it can be transmitted, otherwise the bandwidth required (200 Mbps) would be more than the capacity of an average broadband-supporting DSL line (max. 8–10 Mbps). In addition, compression algorithms also help in error correction at the receiver end. However, compression reduces the quality of the picture to a certain extent. Some of the key video compression technologies are discussed below.

9.16.1. MPEG2

MPEG2 (Moving Picture Experts Group) is hardware-based technology and is currently the most widely used video broadcast technology standard. It reduces the bandwidth required for transmitting a video signal from 200 Mbps to 3 Mbps. However, MPEG2 technology is not capable of supporting IPTV, which requires per-channel bandwidth to be less than 2 Mbps.

9.16.2. MPEG4

MPEG4, a software-based technology is the latest compression standard developed by MPEG. It requires only 1.6 Mbps or less to deliver broadcast-quality digital video. It also includes advanced interactivity that enables the user to stream interactive content. MPEG4 technology is fully capable of supporting IPTV transmission.

9.17. Enterprise IP Communication Service

So far, most of the discussion in this chapter has been on how IP technology is helping residential customers. However, enterprises also benefit enormously from IP-based services. Around the world, enterprises are increasingly under pressure to cut down costs and increase employee productivity. Although telephone, the Internet, PBX and other communication technologies have cut costs and increased productivity, newer IP based communication tools promise more. PBX or Centrex are good but cumbersome, expensive, and have very few features. In addition, separate networks are required for voice, video and data. In contrast, IP-based communication systems provide all services over a single network. Users can check email and voice mail, video conference, receive and make calls from any device. Customer contact centers will be far more dynamic, with features like automatic customer look-up, and integrated IVR. IP-based systems also provide in-house and higher administrative control, allowing enterprises to change, add and delete service configuration dynamically and make optimal use of resources. Thus, IP-based communication services not only save costs, but also provide far more enhanced features that can improve employee productivity and customer satisfaction.

Research conducted at enterprises where IP based communication systems are used show at least 10% productivity improvement and up to 50% cost reduction. Enterprises have realized significant savings in day-to-day operations like additions, moves, changes, and in other system administration activities like call forwarding, concierge services, voice mail setup, premises equipment configuration, and end-user support. Enterprises are also able to eliminate long-distance communication costs and save on network maintenance due to using a single network.

For enterprises, there are two types of IP services available in the market today:

*   IP Centrex
*   Hosted IP PBX

9.17.1. IP Centrex

Centrex, or central office exchange service, is an enterprise communication solution where all the switching equipment is located at the central office. The equipment is owned, maintained and operated by the service provider. This frees up the enterprise (especially small and medium-sized enterprises) from managing a complicated infrastructure.
IP Centrex is the IP version of Centrex (Chapter 4). It offers typical IP benefits like increased capacity, integrated network, higher data communication speed, and reduced investment, and far

richer features like unified messaging, Web-retrievable voice mail, click-to-dial, and the integration of directories and calendars with telephony features.

9.17.2. Hosted IP PBX

Private branch exchange (PBX – Chapter 4) is an enterprise communication solution where all the switching equipment is owned by the enterprise and located at their premises. PBX simplifies communications by providing far more control, as the enterprise can add, update, change and delete connections by themselves. In addition, PBX provides far more features than Centrex.

IP PBX is the IP version of traditional PBX. Unlike traditional PBXs that were owned and managed by the enterprise, IP PBX is now offered by providers as a hosted service, referred to as Hosted IP PBX. The equipment is owned by the provider but located at the user's location, allowing far greater control over day-to-day operations.

The obvious question that comes to mind is, which is better, IP Centrex or hosted IP PBX? The answer is that neither is better for all types of enterprises. IP Centrex is good for small enterprises that do not have a dedicated team to deal with the day-to-day job of maintaining a complex communication system. On the other hand, large enterprises that can afford a dedicated team find it cost-effective to use an IP PBX system.

9.18. IP @ 2010

IP is fast becoming the universal vehicle of choice for communications, media and entertainment. Telecommunication companies around the world are replacing their existing analog network with an all-digital IP-based network. Even though IP has been in use for nearly 30 years now, it is only in the last few years that it has suddenly shot to prominence. DSL and VoIP started the trend, and today there are other services like IPTV, IP Centrex, IP PBX, mobile data, mobile TV, cable over IP, and wireless LAN over IP. However, the true significance of IP technology is that it is forming the core of the next-generation networks (NGNs), facilitating affordable triple-play business models that seamlessly integrate voice, video and data. In addition, IP technology is ideally suited to supporting converged services. Already around 50% of all global telecommunications traffic is done over IP, and this will increase to 75% by 2010. IP is the underlying factor for the rapid changes and development happening in the telecommunication industry. It will continue to play a major role in digitization of every aspect of human life, regardless of whether it is business, personal or entertainment.

10. Telecommunication Business Processes

A business process is a collection of related activities executed in a structured sequence to produce a specific service or product for a customer. Some examples of business processes in telecommunications are sales, ordering, billing, provisioning, and trouble management.

A business process is therefore a set of work activities that has:

- A beginning and an end
- Specific inputs and outputs
- A predefined sequence of steps to be followed
- Uses resources

Business processes define the tasks, rules, people and applications involved in delivering goods, services or information to the internal and external customers of the organization. They place strong emphasis upon how an activity is carried out within an organization rather than just the output, usually involve more than one organizational unit, and may be dependent upon one or more other business processes.

Fig. 10.1 depicts a high-level business process that is carried out while selling a product. The process starts by receiving a request from a customer, and ends by informing the customer of the successful completion of the order. In between these two steps, the business process touches upon various organizations like IT, fleet management, accounts, and others. Multiple organizations have to collaborate to deliver the final product to the customer, and processes within each unit can last from a few seconds to several days.

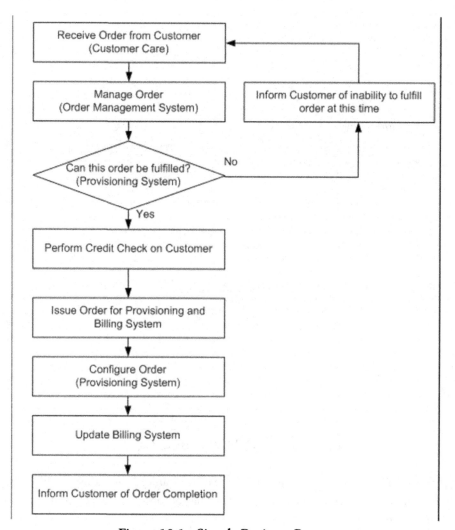

Figure 10.1 - Simple Business Process

10.1. eTOM®

Business processes involved in running a telecommunications company are quite complicated. Fortunately, Tele Management Forum (www.tmforum.org), an international consortium of over 450 major service providers, suppliers and consulting companies, has standardized business processes as part of the enhanced Telecom Operations (eTOM) initiative. eTOM is focused upon creating an industry standard blueprint for the processes to be followed in running a telecommunications organization. For example, there are processes defined that describe how to fulfill an order placed by a customer, processes for managing sales, marketing, product development, partner management, trouble management, billing, and so on.

The eTOM processes are divided into three major categories:

- Strategy, Infrastructure and Product (SIP)
- Operations
- Enterprise management

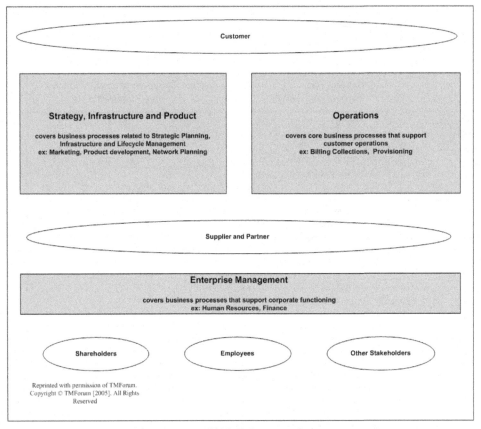

Figure 10.2 - eTOM® Process Categories

Today, planners, managers and strategists in the telecommunications industry widely use eTOM to formulate business processes for their companies. It has become the de facto standard for defining processes to run the industry.

10.2. Benefits of eTOM

Prior to eTOM, every telecommunications company developed and implemented its own business processes. Thus, every company spent an enormous amount of money on defining how to run their business. This meant that every company reinvented the wheel, which led to delays and high operational costs. In addition, commercially available software had to be heavily customized to suit the unique processes, which added to the cost and delays. Partnerships and alliances with other players became a problem as companies followed different ways of doing the same thing.

eTOM provides many benefits to all participants in the telecommunications business. It provides a standard structure, terminology, and classification scheme for describing business processes, and is the blueprint of business processes from which all providers and suppliers work. It provides a common definition for the terms used in the processes and a general description of the activities within each process. A common definition of terms and processes makes it easier for providers to negotiate with customers, suppliers, and other service providers. Other benefits include reduced training costs, clear directions for long-term strategy, and support for a rapidly evolving product portfolio.

10.3. Understanding eTOM

There are literally hundreds of processes followed by a telecommunications service provider. With so many processes, it is hard to understand which one is used where, when, and how. eTOM has a novel way of categorizing these processes into logical groups, which makes it easy to understand where a process falls in the overall set, and when and how it is used. The eTOM framework defines business processes within a series of hierarchical groupings called "levels." Fig. 10.2 shows level 0, which is the base categorization of processes based on three major activities that any industry performs:

- Operations – This grouping covers the core operations that directly supports customer interaction. It includes day-to-day activities or operations like CRM, service management, resource management, and supplier-partner relationship management
- Strategy, Infrastructure, and Product planning – This grouping covers strategic planning activities around marketing, services, product, infrastructure development, and supply-chain management
- Enterprise management – This grouping covers activities that are needed to support the business itself. It includes processes for managing human resources, stakeholder relationships, knowledge, research, and company finances

Each level 0 is further divided into a level 1 group of processes. For the ease of understanding, the level 1 group of processes is categorized into horizontal and vertical process groups:

- Vertical – This group of processes represents a view of end-to-end processes within the business (Fig. 10.3). All processes can be grouped into one of the following vertical categories:

    o   Strategy and commit
    o   Infrastructure lifecycle management
    o   Product lifecycle management
    o   Operations support and readiness
    o   Fulfillment
    o   Assurance
    o   Billing

For example, fulfillment is a vertical grouping of all the activities required in fulfilling a customer's request. As you can see in Fig. 10.4, the vertical and horizontal sub-processes cross each other. For example, CRM activities do not end after fulfilling the order (fulfillment), but extend into assurance and billing for the services rendered.

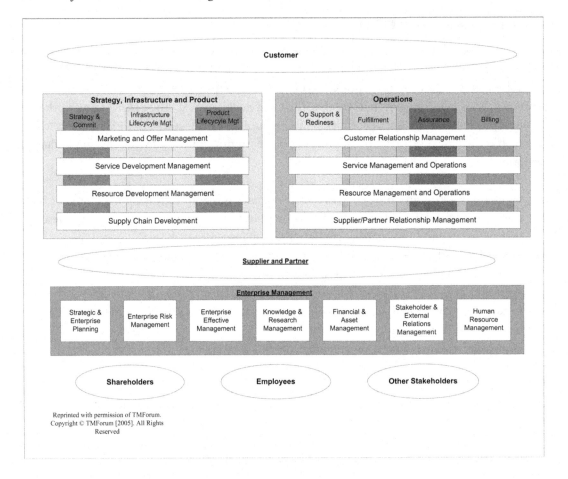

Figure 10.3 - eTOM® Vertical level 1 Process Groups

- Horizontal – This group of processes represents a view of functionally related processes within the business (Fig. 10.4). All processes can be grouped into one of the following horizontal categories:

  o Marketing and offer management
  o Service development management
  o Resource development management
  o Supply-chain development
  o Customer relationship management
  o Service management and operations
  o Resource management and operations
  o Supplier/partner relationship management

For example, customer relationship management (CRM) is a horizontal group that encompasses all CRM-related activities like selling, order handling, problem handling, billing, and collections management. Other horizontal sub-processes are service management and operations, resource management and operations, and supplier/partner relationship management.

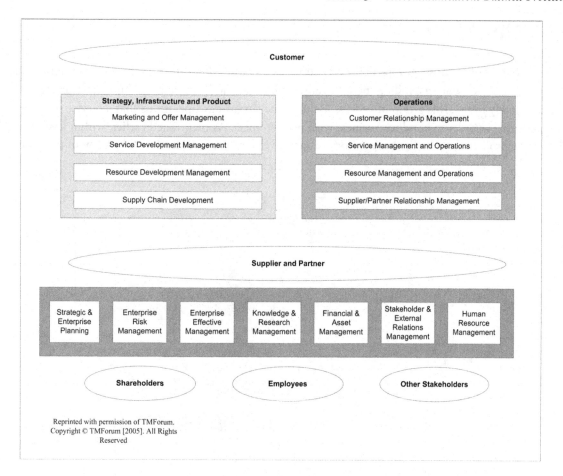

Figure 10.4 - eTOM® Horizontal Level 1 Process Groups

The next three chapters are dedicated to each of the level 0 processes: operations; strategy, infrastructure and product; and enterprise management. Each chapter will delve into the level 1s and briefly discuss the flows.

11. Operations

The operations process group (Fig. 11.1) contains processes that directly support customer interactions like taking orders, billing inquiries/adjustments, service provisioning, problem management, and many more. This group of process includes the day-to-day operations that a service provider performs in order to conduct business with its customers. As mentioned earlier, within level 0, each of the processes can be further categorized into level 1 vertical and horizontal groups of processes.

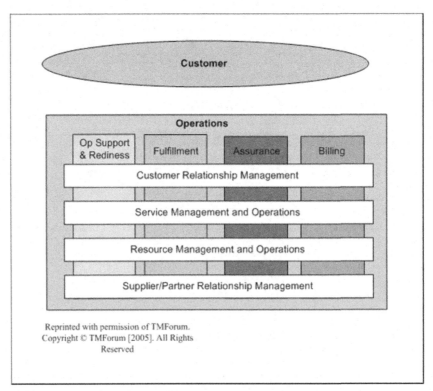

Figure 11.1 - eTOM® Operations Process Groups

11.1. Vertical Processes

The four vertical or end-to-end groups of business processes followed in day-to-day operations are:

*   Fulfillment
*   Assurance
*   Billing
*   Operations support and readiness

11.1.1. Fulfillment

This group of processes is responsible for fulfilling a customer's request for products and services in a timely and correct manner. Key activities are selling, handling customer orders, configuring, and activating services as requested. In general, activities in this group are responsible for successfully fulfilling a customer's requirements using the service provider and partner's infrastructure.

11.1.2. Assurance

This group of processes is responsible for assuring that the services sold are up and running as promised. Key activities are accepting calls reporting trouble, opening trouble tickets, resolving the trouble, resuming service, and informing the customer of the successful handling of trouble. This group of processes also includes managing performance issues. A service might be up and running but may not be performing as promised. Activities in this group frequently detect and correct problems without the customer ever knowing about them. In general, activities in this group perform proactive and reactive maintenance to ensure that services provided to the customer are continuously available and performing as promised through service level agreements (SLA).

11.1.3. Billing

This group of processes is responsible for all billing-related activities. Key activities are collecting and processing usage data, creating bills, managing customer inquiries about their bills, adjusting bills, processing payments, and so on. This group also provides information about the rating and taxation of products.

11.1.4. Operations Support and Readiness

This group of processes is responsible for supporting the day-to-day fulfillment, assurance and billing processes. These processes are not real-time and do not interface with the customer, but are back-office processes that prepare the enterprise to interact with the customer, take care of their requirements, and support day-to-day operations.

11.2. Horizontal Processes

The four key horizontal or functional groups of business processes followed in day-to-day operations are:

- Customer relationship management
- Service management and operations
- Resource management and operations
- Supplier/partner relationship management

11.2.1. Customer Relationship Management

This group of processes is responsible for acquiring, servicing and retaining customers. Key activities are selling, handling customer orders, problem management, and billing and collections management. In addition, this group of processes also includes managing knowledge about customer requirements, likes and dislikes based on criteria like age demographics, opinions about the service they are receiving, personalizing service to meet their needs, and so on.

11.2.2. Service Management and Operations

This group of processes is responsible for the running of services. A service is the actual product that a customer uses, such as a wireline voice, cellular line, T1 or DSL. All the knowledge required to run a service (access, connectivity and content) lies within this group. It is responsible for making sure that the services sold to a customer are up and running. Key activities within this group of processes are to take care of configuring and activating new services, handling any day-to-day problems, and making sure the quality of the currently running services are as promised. In addition, short-term capacity planning and initiatives to improve service performance and manage costs are also undertaken by this group of processes.

11.2.3. Resource Management and Operations

This group of processes is responsible for configuring, activating, troubleshooting and managing the performance of network infrastructure resources like switches, routers, fiber and cable. Other resources involved in delivering services, like software applications and servers, are also managed by this group.

11.2.4. Supplier and Partner Relationship Management

This group of processes is responsible for managing interactions with suppliers and partners. It is a key activity group because suppliers supply all of the resources (switches, routers, line cards) required to provision services. In addition, almost all service providers sell partner products all the time. Therefore, the success of a service provider largely depends upon having a streamlined interaction with suppliers and partners.

The existence of S/PRM processes enables a direct interface with the appropriate processes of suppliers and/or partners. Typical processes include issuing purchase orders, tracking them, handling problems, billing purchases, and authorizing payment.

12. Strategy, Infrastructure and Product

The strategy, infrastructure and product (SIP) group of processes (Fig. 12.1) is involved in developing long-term strategies in the areas of marketing, developing new services, resource planning to support growth, and finally developing relationships with suppliers and partners. This group of processes also deals with the lifecycle management of infrastructure and product. Infrastructure lifecycle management deals with the development and deployment of new infrastructure, assessing the performance of the infrastructure, and taking action to meet performance commitments. Product lifecycle management deals with introducing new products in the form of services delivered to customers, and assessing and taking action on product performance.

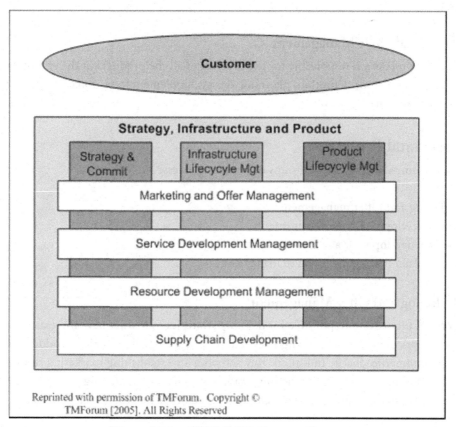

Figure 12.1 - eTOM® SIP Process Groups

12.1. Vertical Processes

The three vertical or end-to-end groups of business processes in SIP are:

- Strategy and commit
- Infrastructure lifecycle management
- Product lifecycle management

12.1.1. Strategy and Commit

This group of processes is responsible for creating enterprise-wide strategies for marketing and offer management, service development, resource development, and supply-chain management. Key activities are gathering and analyzing data from various quarters, coming up with a strategy based on the results of the analysis, and finally planning the execution of the strategy. This group of processes also tracks the success and effectiveness of the strategies currently being implemented, making adjustments as required.

12.1.2. Infrastructure Lifecycle Management

This group of processes is responsible for the development and deployment of new infrastructure (IT, network and operational), assessing performance of the existing infrastructure, and taking action to meet quality and performance commitments for services sold. Key activities are identifying new requirements, exploring new possibilities, and finally designing, developing and implementing infrastructure that will support the goals of marketing, services, and resource and supply-chain management.

12.1.3. Product Lifecycle Management

This group of processes is responsible for the lifecycle of all the products in the enterprise's portfolio. Key activities include definition, planning, design, implementation, forecasting, retirement, and reporting performance.

12.2. Horizontal Processes

The four horizontal or functional groups of business processes in SIP are:

- Marketing and Offer management
- Service development and management
- Resource development and management
- Supply-chain development and management

12.2.1. Marketing and Offer Management

This group of processes is responsible for defining strategies to market new and existing products and services, pricing them, identifying sales channels, assessing and tracking product performance, analyzing customer feedback, and so on. Key activities are establishing marketing strategy, establishing product and services portfolios, defining capability, creating marketing infrastructure, defining promotion strategy, campaign management, and launching new products and services.

12.2.2. Service Development and Management

This group of processes is responsible for planning, developing and delivering services that the provider can sell. Key activities are developing new services, assessing the performance of existing services, and deploying and providing support to operations.

12.2.3. Resource Development and Management

This group of processes is responsible for planning, developing and delivering the resources needed to support the services sold by a provider. It includes functionalities necessary for defining the strategies for the development of the network and other physical and non-physical resources, the introduction of new technologies and working with existing technologies, managing and assessing the performance of existing resources, and ensuring that capabilities are in place to meet future service needs.

12.2.4. Supply-Chain Development and Management

This group of processes is responsible for managing interactions with suppliers and partners who are involved in maintaining the supply chain. The supply chain is a complex network of relationships that a service provider manages in order to source and deliver products. In the e-business world, companies are increasingly working together with suppliers and partners in order to broaden the products they offer and improve their productivity. These processes ensure that the best suppliers and partners are chosen, and that an infrastructure is in place to enable the smooth flow of information.

13. Enterprise Management

The enterprise management group of processes (Fig. 13.1) includes basic business processes that are required to run any business. This group focuses upon running the company and interacts with almost every other process in the enterprise, including operations and SIP processes. Enterprise management includes processes for financial management, human resource management, legal management, regulatory management, etc. It also sets corporate strategies and directions, and provide guidelines and targets for the rest of the business. In many ways, it is more appropriate to call these functions rather than processes.

13.1. Processes

Some of the key business processes in enterprise management are:

- Strategic and Enterprise planning
- Enterprise Risk management
- Financial and Asset management
- Stakeholder and External relations management
- Human resource management

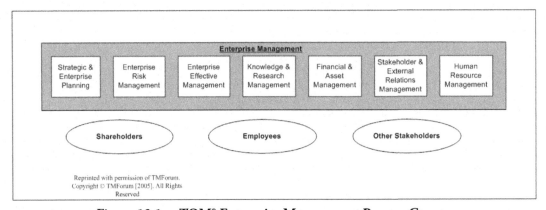

Figure 13.1 - eTOM® Enterprise Management Process Groups

13.1.1. Strategic and Enterprise Planning

This group of processes focuses upon developing strategies and plans that determine the business direction and focus of the enterprise. They address which markets to cater for, the financials, mergers and acquisitions, coordinating between different business units, diversification, outsourcing, investing in new markets, and a host of other strategic issues. They develop the vision for the enterprise – and the overall plan to accomplish that vision.

Functionally, the processes in this group can be divided into four areas:

- Strategic business planning – These processes provide strategic direction to the business.
- Business development – These processes deal with expanding business through new investment, diversification, and mergers and acquisitions.
- Enterprise architecture management – These processes deal with defining, developing, main-

taining and managing architecture for the enterprise, and providing a blueprint for the IT and network operating environment.

- Group enterprise management – These processes plan and manage coordination across business units and subsidiaries.

Strategic and enterprise planning processes are executed by the highest echelons of power in the enterprise.

13.1.2. Enterprise Risk Management

This group of processes focuses upon assuring that the risks and threats to the enterprise are identified, and that appropriate controls are in place in order to minimize or eliminate any risks. Successful risk management ensures that the enterprise can support its mission-critical operations, processes, applications and communications, even in the face of a serious incident, security threats/violation, and fraud attempts.

Functionally, the processes in this group can be divided into five areas:

- Business continuity management – These processes develop strategies, policies, plans, organizational roles and responsibilities, and escalation procedures for ensuring the continuation of business as usual in the event of an interruption.
- Security management – These processes are responsible for security-related corporate policies, guidelines, best practices, and auditing compliances.
- Fraud management – These processes are responsible for corporate policies, guidelines, best practices, and control procedures to deal with fraud.
- Audit management – These processes are responsible for ensuring that standard operational procedures and controls to manage risk are in place.
- Insurance management – These processes are responsible for assessing and managing assets and activities within the enterprise where risk aspects are insurable, and analyzing the costs/benefits of undertaking specific insurance.

13.1.3. Financial and Asset Management

This group of processes is responsible for managing the finances and assets of the enterprise.

Functionally, the processes in this grouping can be divided into three areas:

- Financial management – These processes are responsible for general ledger, accounts payable, accounts receivable, expense reporting, revenue assurance, payroll, book closing, tax planning, financial statements, etc. The processes are accountable for the financial health of the enterprise, managing cash flow, and auditing for compliance to financial and expense policies.
- Asset management – These processes are responsible for all financial and policy aspects of the physical assets (building, fleets, consumables, infrastructure, inventory) of the enterprise.
- Procurement management – These processes are responsible for policies regarding corporate procurement like purchasing, receiving and returning goods, warehousing, and transport and physical resource distribution.

13.1.4. Stakeholder and External Relations Management

This group of processes is responsible for managing the enterprise's relationship with internal and external stakeholders like shareholders, employee organizations, regulators, and the local community.

Functionally, the processes in this group can be divided into five areas:

- Brand management and advertising – Brand management processes set the overall policies and strategies for the management of the enterprise's brands. It addresses ways of building and maintaining the brand, and issues related to product naming, use of brand, co-branding, etc. Advertising processes develop and execute advertising strategies in support of the overall enterprise, business unit, and specific products. They are used to target customer segments, create relevant advertising programs, execute these programs, and evaluate the effectiveness of these programs once they are run.
- Shareholder relations management – These processes manage the relationship between the enterprise and its shareholders, consistent with all business, financial, legal and regulatory requirements.
- Legal management – These processes are responsible for ensuring that the enterprise complies with all relevant legal requirements. They are also responsible for carrying out legal requests within the enterprise, supporting the enterprise by providing legal advice related to business decisions, and proactively notifying the enterprise of relevant changes or trends that can affect the legal environment. These processes deal with any legal action taken on behalf of or against the enterprise.
- Regulatory management – These processes ensure that the enterprise complies with all existing government regulations. They are responsible for legally influencing pending regulations and statutes to benefit the enterprise and to inform the enterprise of the potential consequences of pending legislation or regulations. In addition, these processes are responsible for tariff filings as required.
- Community relations management – These processes are responsible for conducting and/or participating in local activities, providing financial and other support to schools, libraries and hospitals in the communities that the enterprise serves. The processes also deal with public relations and handle contact with customer interest groups and community groups.

13.1.5. Human Resource Management

This group of processes is responsible for managing the human resources working for the enterprise. They deal with hiring and firing employees, salary structure, performance appraisal, benefits programs, and organization structure.

Functionally, the processes in this group can be divided into four areas:

- HR policies and practices – These processes are responsible for performance appraisal, benefits, occupational health and safety, equal employment opportunity, compensation guidelines, code of conduct, hiring and termination guidelines, employee satisfaction measurement and management, etc.
- Workforce strategy – These processes are responsible for understanding the human resource requirements of the business and defining the competencies and skills required to satisfy these

needs. The processes create the strategies needed to ensure that the correct type, quantity and quality of employees are available in the right locations for future business. They work with all areas of the enterprise to devise any changes required in the workforce to support the overall strategy of the enterprise.

- Workforce development – These processes are responsible for the development of employees in order to meet the needs of the business. They perform competency modeling, skills assessment, job and employee strength profiling, succession planning, training development and delivery, career development, work design, employee recruitment, etc.
- Employee and labor relations management – These processes are responsible for managing relationships with employees and employee groups. They include the definition of terms of employment, labor contract development, union contract negotiations, and arbitration management. In addition, they interface with employee groups, manage counseling programs, employee involvement in the community, and charities in the name of the company, etc.

Section 4 – Telecommunications Standards and Regulations

14. Standards

Standards are industry-wide accepted definitions of protocols, data formats, and interfaces. They help interoperability among different communication systems used around the world, and facilitate the easy integration of the different components in a communication system.

14.1. Why are Standards required?

During the early days of telecommunications, communication between two provinces or countries was impossible as they were served by different companies who did not follow the same standards. The formats, technology and systems used were different, and this meant that messages had to be transcribed, translated and handed over at frontiers, then re-transmitted over the network of the neighboring country.

As the telecommunication networks within each country in Europe and the US grew rapidly, it was clear that this mode of operation was not efficient and was actually hampering growth. It finally prompted 20 European states to develop a framework agreement covering international interconnection. At the same time, the group decided upon common protocols to standardize equipment and create uniform operating instructions, a common international tariff, and accounting rules, all to facilitate international interconnection.

Standards have evolved quite a bit since then and have become an extremely important part of the telecommunications business. Their main goal is to provide:

*   Interoperability – This is the primary reason for standards. Telecommunications business is technically very complicated, with hundreds of components talking to each other. Unless there are standards for all vendors to follow the same specification in manufacturing these components, they are unlikely to work in harmony
*   Quality assurance – Minimum performance expectations are set while standards are defined
*   Uniform evolution – Standards create a roadmap on how the technology will evolve so that equipment manufacturers and providers can evolve accordingly

Today, standards have become pseudo-laws that are followed by manufacturers and service providers. There are standards for network equipment, softwares, protocols, etc., developed by an international community of network designers, operators, vendors and researchers working under the auspices of various international standards bodies like the International Telecommunication Union (ITU) and the European Telecommunications Standards Institute (ETSI).

This chapter takes a look at several international standards bodies. They are classified based on the focus of their operations.

14.2. Standards Bodies

14.2.1. ITU

The International Telecommunication Union (ITU) is the telecommunication standards body of the United Nations. It is unique among international organizations in that it was founded on the principle of cooperation between governments and the private sector. ITU's activities, policies, and strategic direction are determined and shaped by the industry it serves. Members of this body are policy-makers and regulators, network operators, equipment manufacturers, hardware and software developers, regional standards-making organizations, and financing institutions.

The activities of ITU cover all aspects of telecommunication, from setting standards that facilitate the seamless interworking of equipment and systems on a global basis, to adopting operational procedures for the vast and growing array of wireless services. Experts drawn from leading telecommunication organizations worldwide carry out the technical work, preparing the technical specifications for telecommunication systems, networks and services, including their operation, performance and maintenance. Their work also covers the tariff principles and accounting methods used to provide international service. All ITU recommendations are non-binding voluntary agreements.

14.2.2. ATIS

The Alliance for Telecommunications Industry Solutions (ATIS) is a United States-based body that is committed to rapidly developing and promoting technical and operations standards for the communications and related IT industry worldwide.

Over 1,100 industry professionals from more than 350 communications companies actively participate in its 22 industry committees and incubator solutions programs. ATIS develops standards and solutions addressing a wide range of industry issues in a manner that allocates and coordinates industry resources and produces the greatest return for communications companies. It creates solutions that support the rollout of new products and services into the communications marketplace, and its standardization activities for wireless and wireline networks include interconnection standards, number portability, improved data transmission, Internet telephony, toll-free access, telecommunication fraud, and order and billing issues, among others.

14.2.3. ETSI

The European Telecommunications Standards Institute (ETSI) is an independent, non-profit organization, whose mission is to produce telecommunication standards for today and for the future.

Based in Sophia Antipolis, France, it is officially responsible for the standardization of information and communication technologies (ICT) within Europe. These technologies include telecommunications, broadcasting and related areas such as intelligent transportation and medical electronics. It unites 688 members from 55 countries inside and outside Europe, including manufacturers, network operators, administration, service providers, research bodies, and users. It plays a major role in developing a wide range of standards and other technical documentation as Europe's contribution to worldwide ICT standardization. Its prime objective is to support global harmonization by providing a forum in which all the key players can contribute actively. It is officially recognized by the European Commission and the EFTA secretariat.

14.2.4. GSMA

Founded in 1987, the Global System for Mobile Communications Association (GSMA) is the global trade association that exists to promote, protect, and enhance the interests of GSM mobile operators throughout the world. As of April 2007, it consisted of more than 700 second and third-generation mobile operators and more than 200 manufacturers and suppliers. GSM is a living and evolving wireless communications standard that already offers an extensive and feature-rich family of voice and data services. The GSM family consists of today's GSM, general packet radio service (GPRS), enhanced data rates for GSM evolution (EDGE), and third-generation GSM services (3GSM) based on the latest WCDMA technology.

14.2.5. 3GPP

The 3rd Generation Partnership Project (3GPP) is a collaboration agreement that was established in December 1998 to develop a suite of 2.5G to 3G wireless systems derived from GSM. The collaboration agreement brings together a number of telecommunications standards bodies like ARIB, CCSA, ETSI, ATIS, TTA and TTC.

The original scope of 3GPP was to produce globally applicable technical specifications and technical reports for a third-generation mobile system based on GSM. The scope was subsequently amended to include evolved radio access technologies like GPRS and EDGE.

14.2.6. CDG

The CDMA Development Group (CDG), founded in 1993, is an international consortium of companies focused upon adopting and evolving CDMA wireless systems. It comprises CDMA service providers and manufacturers and seeks to ensure interoperability among systems and the availability of CDMA technology to consumers. Other activities include requirement definition, promotion, facilitation, cooperation, minimizing time-to-market, and creating global economies of scale.

14.2.7. 3GPP2

3GPP2 is CDMA's answer to competing technology GSM's 3GPP forum. The mandate is to develop a suite of 3G wireless systems derived from CDMA technology.

14.2.8. W3C

The World Wide Web Consortium (W3C) is an international consortium where member organizations, a full-time staff, and the public work together to develop standards for the Web. Its mission is "To lead the World Wide Web to its full potential by developing protocols and guidelines that ensure long-term growth for the Web."

W3C's global initiatives also include nurturing liaisons with national, regional and international organizations around the globe. These contacts help W3C maintain a culture of global participation in the development of the World Wide Web. W3C coordinates particularly closely with other organizations that are developing standards for the Web or Internet in order to enable clear progress.

14.2.9. IETF

The Internet Engineering Task Force (IETF) is a large, open international community of network designers, operators, vendors and researchers concerned with the evolution of Internet architecture and the smooth operation of the Internet.

The mission of IETF is to produce high quality, relevant technical and engineering documents that influence the way people design, use and manage the Internet in such a way as to make it work better. These documents include protocol standards, best current practices, and informational documents of various kinds.

14.2.10. ICANN

The Internet Corporation for Assigned Names and Numbers (ICANN) is an internationally organized, non-profit corporation that has responsibility for Internet protocol (IP) address-space allocation, protocol identifier assignment, generic and country codes, top-level domain name system management, and root server system management functions.

As a private-public partnership, ICANN is dedicated to preserving the operational stability of the Internet, promoting competition, achieving broad representation of global Internet communities, and developing policy appropriate to its mission through bottom-up, consensus-based processes.

14.2.11. DSL Forum

DSL Forum is a consortium of approximately 200 leading industry players covering telecommunications, equipment, computing, networking, and service provider companies. Established in 1994, the forum continues its drive to develop the full potential of DSL in order to meet the broadband needs of the mass market. In 14 years, the DSL Forum has moved through defining the core DSL technology to establishing advanced architecture standards and maximizing effectiveness in deployment, reach, and application support. It has driven the global standardization of ADSL, SHDSL, VDSL, ADSL2plus and VDSL2, and more are in progress. These will provide a complete portfolio of DSL technologies designed to deliver ubiquitous broadband services for a wide range of situations and applications that will continue the transformation of our day-to-day lives in an online world.

14.2.12. IPCC

The International Packet Communication Consortium (IPCC) is a leading international industry association dedicated to accelerating the deployment of VoIP, video over IP, and packet technologies and services over converged wireless, wireline, and cable broadband networks. The association comprises service providers, solution providers, systems integrators, and government agencies translating industry standards into revenue-generating services. It is involved in developing technical frameworks for converged services in wireline, cable, 3G, Wi-Fi and WiMax networks.

14.2.13. MFA Forum

The MPLS, Frame Relay and ATM (MFA) Forum is an international, industry-wide, non-profit association of telecommunications, networking, and other companies focused upon advancing the deployment of multi-vendor, multi-service, packet-based networks, associated applications, and interworking solutions. The goal of the forum is to create specification on how to build and deliver MPLS, frame relay and ATM networks and services.

14.2.14. CableLabs

Founded in 1988 by members of the cable television industry, Cable Television Laboratories, Inc. (CableLabs) is a non-profit research and development consortium dedicated to pursuing new cable telecommunications technologies and helping its cable operator members integrate those technical advancements into their business objectives. It seeks to enable interoperability among different cable systems, facilitating the retail availability of cable modems and advanced services, and helping cable operators deploy innovative broadband technologies.

14.2.15. DLNA

The Digital Living Network Alliance (DLNA) is a cross-industry organization of leading consumer electronics, computing industry, and mobile device companies. They share a vision of a wired and wireless network of interoperable consumer electronics, personal computers, and mobile devices in the home and on the road, enabling a seamless environment for sharing and developing new digital media and content services. DLNA is focused upon delivering interoperability guidelines based on open industry standards in order to complete cross-industry digital convergence.

DLNA has published a common set of industry design guidelines that allow manufacturers to participate in a growing marketplace of networked devices, leading to more innovation, simplicity, and value for consumers.

14.3. Are there too many Standards?

By last count, there were more than 300 organizations working on standards from wireless to VoIP and IPTV. As a senior executive from ATT rightly said, "Too many standards mean no standards!" To compound this problem, some are duplicate initiatives and some are competing, which makes them difficult to implement. On the other hand, it is important to remember that telecommunications is a vast field and many areas need standardization.

15. Regulations in the Telecommunications Industry

Up until almost the last decade of the twentieth century, the telecommunications business in almost all countries around the world was either run by monopolistic private companies or the state. There was not much growth because neither of the models fostered innovation and entrepreneurship - the basic criteria required for rapid growth. Thus, the growth in telecommunications industry was stagnating and existing operators lacked ability to serve the tremendous demand for services. By early 1990s, it was clear to many governments around the world that private sector capital was needed to upgrade the networks and introduce innovation into the market. They started overhauling existing archaic and restrictive regulations. In its place, new pro-competitive regulations that supported growth, private investment and entrepreneurship were introduced. Numerous state-owned telecommunication companies were privatized and entrepreneurs started building new ones to service the huge demand. This was the start of a new beginning in the history of telecommunications industry.

15.1. Goals

In theory, market forces are supposed to generate competition and only allow the fittest to survive and flourish, but this does not occur in the telecommunications business. The very nature of the telecommunications business requires an enormous amount of capital over a very long period. The barrier to entry for newcomers is very high, and the probability of newcomers making it on their own is very low. As such, regulatory bodies have been set up by almost every country to sustain a competitive environment and foster growth in telecommunication industry. The primary goals of the regulatory bodies are:

- Nurture competition in the market – As mentioned earlier, the telecommunications business has a very high barrier for entry in terms of investment and expertise. If left to the market, the well-established and financially strong pre-existing operators would never allow the startups to run. An environment conducive to the growth of new players has to be established in order to transform the telecommunications market from state-owned or monopolies into a competitive one where old and new comers can survive and thrive. Regulatory bodies implement regulations to maintain competition in the market.

- Universal access is a basic right – One of the key social objectives of governments is to provide affordable voice service to all, regardless of who the user is. However, not all consumers and markets are equally profitable. If left to the market, telecommunication companies would only cater to cities and upscale neighborhoods where operations are profitable. Regulatory bodies force telecommunication companies to implement this social policy.

- Prevention of anti-competitive practices – Even in developed countries like the USA, it is quite common to see providers engaging in anti-competitive practices. For example, in many communities, especially smaller ones that are served by only one provider, the provider usually charges more and provides a poor service. Regulatory bodies are required to improve competition.

- Mergers and acquisitions – Mergers and acquisitions are a common phenomenon in telecommunication market. However, many of the times, these activities are undertaken keeping only the best interest of the company and not the communities served. Regulatory bodies are required to ensure that the interests of the communities are given priority as well.

- Rulemaking – The business of telecommunications is constantly changing due to new technologies, which makes it very difficult to create regulations to meet goals. In order to overcome this problem, in most countries, legislative bodies set the general terms and conditions under which telecommunications providers must offer service to the public. The legislative body then allows the regulator to come up with specific regulations that are required from time to time to implement these broad policies. The process by which the regulator adopts such rules and regulations is generally referred to as "rulemaking"
- Adjudication – The process by which the regulator ensures that telecommunication providers comply with its rules and regulations is referred to as enforcement or adjudication. To enforce compliance with rules, the regulator must have the power to investigate the actions and records of all telecommunication providers, and must have the authority to impose sanctions and penalties for the violation of regulations.
- Managing scarce resources – Telecommunication businesses utilize scarce national resources like the radio spectrum, underground ducts for cabling, and land in the public domain to build infrastructure. Regulations are required to manage these resources efficiently and effectively
- Protecting consumers – Regulations are also required to define consumer rights, draft appropriate legislation, educate consumers about their rights, and defend consumers against the anti-competitive behavior of providers. For example, regulations like local number portability have benefited consumers immensely, as they can now switch providers and still keep the same number.

15.2. Top Regulatory Reforms

The following are some of the major reforms introduced by the regulatory bodies around the world and the reasons for doing so:

- Privatization – The telecommunications sector around the world is gradually being opened up to private players. This is mainly to attract much-needed investment in order to expand the reach of basic services and introduce new services into the market. This move also introduces competition, making the industry more efficient
- Licensing – This is an effective way to share the critical resources (especially the radio spectrum and public land) required for telecommunications. The licensing process also ensures that an operator is financially and technically capable of conducting a complex business and that they will always uphold the country's laws
- Price regulation – Price regulation is important for those services that are considered universal - like wireline service, local call, and public telephone booth charges. The prices of universal services have to be based on affordability
- Transparent regulatory process – The key to attract private investment into the telecommunication sector is to demonstrate transparent regulatory process so that investors are assured of impartiality towards the state-owned operator
- Interconnection and unbundling – The telecommunications business requires an enormous amount of investment in order to build the network. It was obvious that it would take private players years to build such a network and compete with the incumbent (existing service providers like BT, ATT, Bell Canada, Telstra, BSNL, Deutsche Telecom, etc). The solution was to force the incumbent to share its network of local loops so that a private operator can use it to sell local services. In addition to this, it was also important to force the incumbent to allow private players' networks to talk to the incumbent's network so that they can sell meaningful services that work in areas covered by the incumbent

- Universal access funds – One of the key reforms adopted by almost all the regulatory bodies is the creation of funds for supporting universal access. These funds support services in areas where providing basic service is not profitable.

15.3. Regulatory Bodies

Regulatory organizations around the world can be classified into three categories:

- National government – It is not that common, but even today in some countries, the federal government functions as the regulatory authority for telecommunications. Policies are made and implemented by the same officials who may also be running the state-owned network operations
- Independent regulatory authority – Almost all developed economies have independent bodies acting as regulatory authorities. Policies that are in the interest of the nation are made by the government, but they are implemented impartially by the independent regulatory body
- International organizations – Many international organizations like ITU (International Telecommunication Union), WTO (World Trade Organization), and AFDB (African Development Bank) also play a major role in regulating the telecommunications business. The focus of these organizations varies; some have a regional or global mandate to improve regulations, while others provide technical assistance or fund activities that improve regulatory expertise and implementation.

15.4. Are Regulations necessary?

Regulations in the telecommunications industry are required to create an environment that balances the need to protect consumers, provide incentives for competition, innovation and investment. Unfortunately, very few regulatory bodies can claim that they have achieved this balance. Many of the policies have stifled growth and slowed investment. For example, the policy of unbundling forces the incumbent to build and maintain a network at enormous cost and then give new providers access to that network at discount prices set by regulatory authorities. The result is that incumbents have slowed down the creation of new networks, as it is no longer profitable. Similarly, the price cap regulation has forced companies to provide services at prices that are below cost in many places, which is not very encouraging either. In addition to all this, almost all regulatory bodies have failed to come forward with clear-cut policies at the right time, leading to confusion and lawsuits. Policy uncertainty erodes market value and stifles capital spending by investors.

In addition to the failure of regulatory bodies in doing their job effectively, another reason for not having a very strong regulatory environment is that the nature of the telecommunications business has completely changed in the last decade. Technology has changed the very landscape of business. Regulatory bodies were set up when there were very few services provided by one or two monopolistic companies, but now the landscape boasts several players taking advantage of new technology like wireless and IP to compete effectively in the market. Many of the regulations today are based on the network used to provide the service. For e.g, there are specific regulations on voice service provided using copper wire, but the same regulations do not apply for voice service over IP. Similarly, there are many regulations for providing TV service over cable, but those regulations do not apply to TV over IP or TV over mobile channel.

Regardless of these point failures, the regulatory bodies world over are doing a good job. The sustained growth seen in every corner of the world stands testimony to that. However, every country must assess its situation and decide what and how much regulation is needed. The assessment must be based on the competition in the market – as competition increases, regulation should decrease; and on the level of implementation of social policy on the ground.

15.5. Challenges

All regulatory bodies around the world face significant challenges in achieving their objectives in privatization, licensing, rulemaking, enforcement, and other areas. However, the biggest challenge of all is convergence. Convergence (chapter 16) is defined as the concept whereby service providers can provide any service using any network. This is because almost all regulations that are out there today are based on the assumption that services and networks are interdependent. For example, regulatory bodies heavily regulate the wireline voice service (based on price, access and coverage) provided over copper wire. However, today, the same voice service can be provided by VoIP, mobile, peer to peer, and many other ways, which are not equally regulated. Thus, the challenge now is whether to remove the regulations on wireline voice services or regulate the other ways that voice service is provided. Also, earlier, there were different laws governing voice, video and data, but now with IP technology the same network is capable of carrying all signals. Thus, all regulations that were built on the earlier assumption that services and networks are interdependent now need to be reexamined. New regulations that span across services and networks need to be written in such a way that they do not stifle the phenomenal growth that the telecommunications world has witnessed over the last few years, but at the same time also ensure that the public policy objectives of the regulatory bodies are met and that the environment is competitive.

15.6. The Way Forward

Changes in telecommunications are happening faster than the abilities of the regulatory authorities to keep up with them. The telecommunications business is growing at an enormous pace. To continue with this upward trend, it is necessary to have universal standards for manufacturing equipment, handsets, protocols, controlling cyber-crime, protecting consumer interests, intellectual property rights, and a host of other public issues, otherwise growth will be stifled due to disparities. It is clear that to keep up with such rapid change, regulatory authorities around the world need to cooperate and collaborate in defining the regulatory requirements that will foster rapid growth. Regulators across the globe must all cooperate and draw up agreements under the auspices of global bodies like ITU. They must keep themselves abreast of changes beyond their borders if they are to remain effective.

Section 5 – Trends in the Marketplace

16. Convergence

Through out the 20th century, the telecommunications industry was clearly divided into sectors based on the service they provided. Wireline companies provided voice and data, but did not provide video service. Cable companies provided only video service and nothing else. Wireless companies could only provide voice service, but no broadband or video service. The primary reason for this was that each service of these services were developed over a period of time. Each of these industries lived separate lives as shown in Fig 16.1.

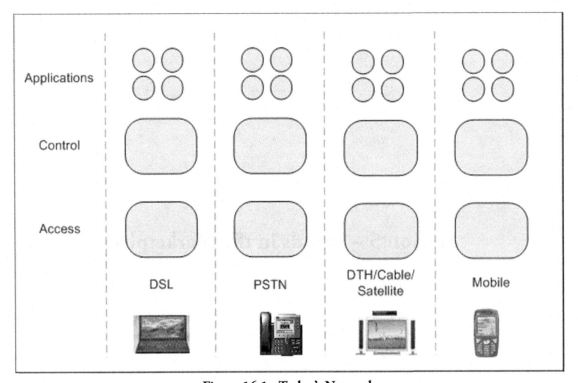

Figure 16.1 - Today's Network

However, the advent of use of Internet Protocol (IP) technology to provide all services using a single protocol (Fig 16.2) started shaking up the industry. It became clear that there was no need to maintain three separate networks, which was becoming expensive and cumbersome to maintain and operate, when a single protocol over a single network could do the same job at much lower costs.

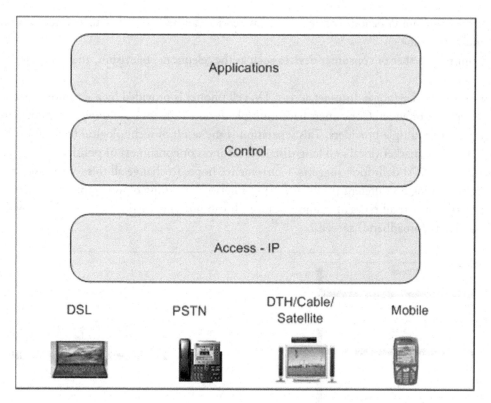

Figure 16.2 - All IP Network

This is dramatically changing the way the industry competes. It used to be that only mobile companies competed with mobile companies, wireline companies competed with wireline companies, and Internet service providers (ISPs) competed with ISPs, but this is no longer the case. Every company, whether fixed, mobile or cable, is trying to provide all the services. It is as if all the players are converging into a single telecommunications industry.

16.1. Definition

The term convergence lacks a single universal definition. It means different things to different players in the industry. Every group, from networks to sales to consumers, defines it in a way that affects them.

The best overall definition probably comes from ITU, it states - "Convergence is technological, market or legal/regulatory capability to integrate across previously separated technologies, markets, or politically defined industry structures."

The European Union (EU) defines convergence as "the ability of different network platforms to carry essentially similar kinds of services, including the coming together of consumer devices such as the telephone, television, and personal computer."

In simple terms, convergence means:

- Transmission of voice, video and data through a single network

- Services are not dependent on network; any service can be accessed from any device, anywhere, and at any time
- Coming together of consumer devices such as the telephone, television, and computer

Today, every service (voice, Internet access, TV, cell phone) is provided by a separate network or even a separate provider. Consumers have multiple devices (one for each service), and multiple accounts with multiple providers. This separation is the result of technological (cellular or PSTN, cable, satellite), market (local and long distance, business or consumer) or political (regulatory) reasons, as the ITU definition suggests. Convergence hopes to change all this – instead of multiple networks running on different protocols, a single network running on one protocol (Internet protocol, IP) will support all types of communication. It can not only carry voice, but also provide video (TV) and data (broadband) as well.

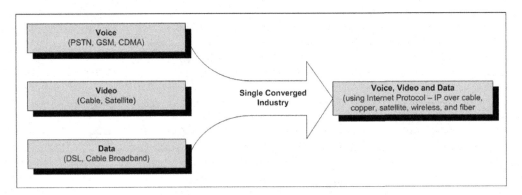

Figure 16.3 - What is Convergence?

The goal of convergence is "any service on any device, anywhere, and at any time." There will be no difference between a cell phone and wireline phone. With a single number, a person can be reached regardless of where they are. Conversation can be seamlessly transferred from a wireless network, to wireline network to Wi-Fi network with the users never knowing that the best possible network has been picked for their communication session. Live sports shows can be viewed on TV at home or transferred to a handheld device while on the move. Voice mails can be checked on a laptop and emails on a mobile phone when necessary. It gets even more interesting with the confluence of entertainment, computing, and communication services, they can be combined to create exciting new products, services, and business opportunities.

16.2. Benefits of Convergence

Convergence has gained so much prominence in the last couple of years because it affords enormous benefits to both providers and consumers. As mentioned above, each service is provided by using different protocol over a separate network. For a provider, it is expensive to maintain multiple networks; difficult, expensive and time-consuming to introduce new services; and very difficult for the network to inter-communicate to realize service convergence. On the other hand, consumers want the costs of telecommunications to go down, and rich service features to be available. This is only possible with convergence.

16.2.1. Benefits to Consumers

- Exciting new converged applications (e.g., caller ID on TV, video on handheld)
- Availability of rich, high-end applications like presence-based, location-based, multimedia SMS, gaming, and many more
- Reduced communication and entertainment costs
- Increased competition in the market, leading to better services

16.2.2. Benefits to Providers

- Single IP-based network (instead of multiple types) that can support all services (voice, video and data)
- Enormous cost savings in operations, equipment, maintenance and staff
- Enormous cost savings due to the efficient use of bandwidth
- Fast and simplified provisioning
- Capability to create innovative services and deploy them rapidly to better meet customer demands. Many of the new multimedia-based applications (e.g., photo share, multimedia SMS, video call) were unthinkable when the networks for voice, video and data were separate
- It is easier to bundle new services into packages, thereby increasing average revenue per user

16.2.3. Benefits to Enterprises

- Overall reduction in the cost of communication-related expenses. This is due to network convergence leading to consolidation of voice, video and data networks into one network, and thereby reducing staff, equipment and maintenance costs.
- New applications possible due to convergence improve employee productivity. Mobile employees have improved access to the office communication network, enabling them to respond to customers, partners and colleagues more quickly
- A wide range of multimedia and multi-service applications, such as video conferencing, instant messaging, collaborative white-boarding, and web-enabled multimedia call centers, helping companies function more effectively
- Single security standards across converged voice and data networks ensure the optimal level of security.

16.3. Frontiers of Convergence

The benefits of convergence are loud and clear. However, achieving a fully convergent industry is a herculean task. Convergence is happening on four key fronts:

16.3.1. Industry Convergence

For years, the communications industry was divided into voice, video and data service silos. They lived and functioned separately. Different technologies were used and there was very little that the industries held in common, except that just about everybody used almost all of them. Digital technology has changed this completely, making it possible for the industries to merge. Today, it is possible to have voice communication over a computer, TV can be watched on a computer or cell phone, and copper wires used for voice communication are now used for broadband Internet access and TV service. In addition, the media and entertainment industries are closely working with the communication industry to create very exciting applications like home banking on TV, user generated news over TV/mobile, mobile ring tones from popular songs, etc. In short, the media and entertainment industry have found a new channel to sell their content.

16.3.2. Service Convergence

Convergence is leading to a combined industry that is capable of providing communication, media and entertainment services. This is leading to a variety of new applications that are formed by combining previously disparate services from each industry vertical. Unified messaging allows you to get your voice mail, email and SMS on any device. Gaming enthusiasts can play games with their friends regardless of where they are. A PVR (personal TV recorder like TiVO) can be controlled from an office computer. Music can be enjoyed on a mobile phone, TV or computer. Service will also become context-aware so that an SMS sent to a person watching TV will be redirected from their cell phone to the TV, rather than just blindly sending it to the cell phone. The possibilities are endless and exciting!.

16.3.3. Device Convergence

The convergence of devices will lead to the ultimate goal of convergence – "one person, one device." A smart device can not only be used for voice communication through any network (fixed, Wi-Fi, cellular, LAN), but also act as a handheld computing device, video and gaming console. High end devices like iTouch, Nokia N95, Video phone, and Xbox are very good examples of multi-capability devices.

16.3.4. Network Convergence

Network convergence is the most time and resource consuming part of migration to a converged industry. The migration is primarily happening from current PSTN, ATM, and Frame Relay to a single IP (run over MPLS) network. BT's 21CN is a good example of this migration.

16.4. Technology Enablers

Without the capabilities provided by the technologies discussed below, convergence would not be possible. Earlier efforts to converge during the 1980s failed mainly because of the lack of the strong technological foundations that are available today. The following are some of the technologies making convergence possible:

16.4.1. Digital Technology

Prior to digital technology, audio, video and text existed in different formats. Audio in analog tapes, text in paper, and video in analog VHS tapes. However, the use of digital technology has converted all of them into a series of zeros and ones. This allows the telecommunication systems to transmit them very easily.

16.4.2. Network Protocol

Two key protocols are also playing a significant role in aiding convergence. Internet protocol (IP) has now made it very easy to transmit any mode of communication (voice, video and data) in the form of packets. The other protocol is MPLS (multi-protocol label service) over which IP runs. Over the years, carriers have maintained one network for each service, but it is clear that maintaining so many different types of networks is not cost-effective. IP over MPLS is the answer to this problem as it is capable of collapsing any type of network into a common IP/MPLS network. The other advantage of IP is that it is capable of running over any type of transport like cable, satellite, copper, fiber and wireless.

16.4.3. IP Multimedia Subsystem (IMS)

One of the key goals of convergence is the ability to get any service over any device, even if the device is on different network. IMS provides a unified architecture that supports a wide range of IP-based services over both packet and circuit-switched networks, employing a range of different wireless and wireline technologies. A user watching a game on TV can simultaneously chat to a friend and send a video clip of a goal from the match. The receiver can then watch the clip on his mobile or TV. A single IMS presence-and-availability engine could track a user's presence and availability across mobile, fixed and broadband networks, or a user could maintain a single integrated contact list for all types of communications. In addition, IMS will also allow users to continue their session even if they move from one type of network/device to another. For e.g. A user's session on cell phone can be seamlessly switch over to fixed network when they are closer to one (home or office).

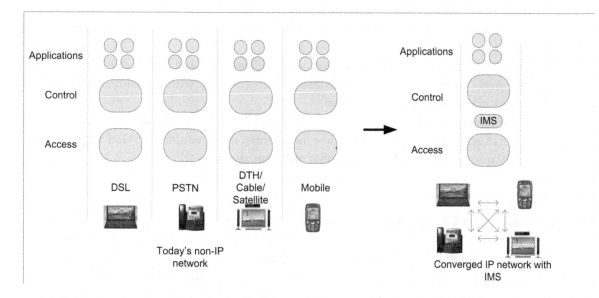

Figure 16.4 - IMS Enabled Converged Network

IMS will have a major impact upon the telecommunications industry. It will lead to a new business model creating tremendous new opportunities for all providers, while at the same time significantly reducing operational costs. It simultaneously lets network owners derive additional value from their networks by opening them up to third parties (including enterprise customers) to develop and offer enhanced and tailored applications of their own.

The basic set of standards for IMS implementation was released in 2004. Both the International Telecommunication Union (ITU) and the European Telecommunications Standards Institute (ETSI) are heavily involved. IMS standards are still being developed to fill in the inevitable gaps and add new capabilities. However, IMS is still untested in real-life within major carrier networks, and its wide scale implementation will not be seen until at least 2010.

16.5. Key Drivers of Convergence

16.5.1. Consumer Demand

Today's telecommunication users are very demanding and well informed about their choices and possibilities. They:

- Expect advanced life style supporting applications over today's plain voice, video and data services
- Expect convergence of communication, media and entertainment services
- Demand higher quality, availability and mobility from applications and services
- Expect to be able to do more with the applications and services for less money
- Expect new applications and services to offer a seamless experience across multiple access technologies, devices and locations, whether wireline or wireless, narrowband, wideband or broadband
- Want a single provider for all their communication and entertainment needs and a single bill too
- Want more control over service features, parameters and applications
- Want applications and services to be personalized (unique service profile), intelligent and context aware
- Want access to the applications and services wherever, whenever and however they want
- Do not want complexity in their applications and services. They want the communications to be simple, manageable and non-intrusive
- Demand communication services that are reliable and safe. One of the major consumer fears of riding all services over IP is that just like the Internet, even traditional services might be affected by spam, malware, adware, and other malicious attacks

16.5.2. Enterprise Demands

Enterprises are always looking for ways to reduce operational expense and run their business more efficiently. In addition to all the demands of an individual customer (listed above), they want to have control and flexibility. With globalization, they need intelligent ways of bridging the distance with smart tools like white-boarding, video conferencing, collaborative working and file sharing. They demand productivity-improving applications so that mobile and remote workers have access to the same tools as in the office, including buddy lists, presence, and stored data.

16.5.3. Service Provider Challenges

Carriers worldwide are facing decreasing average revenue per user, even though usage is increasing. The price of traditional voice, video and data services is declining and margins are thinning rapidly. The competition is intensifying with every service provider providing all services. Cable companies are providing voice services, and telephony-based companies are getting into video business. Wireless operators are offering a competitive service to provide an alternative to voice service from wireline and cable companies. In addition, they all are tying up with each other so that all services can be provided by a single entity. In such a highly competitive market, service providers have to increasingly become customer-centric and pay attention to customer needs if they are to survive and grow. To do this, they have to radically change their business model to reduce churn, increase loyalty, lower operational cost, boost average revenue per user, and respond to the challenges posed by new players like Google and Skype.

16.6. Challenges

Since the concept of convergence is complex and relatively new, the industry is still struggling to get to grips with it. The following are some of the major challenges seen on the ground:

- Cost – Convergence requires an enormous amount of investment in almost every aspect of service delivery. It requires a new network, new support systems, new end-user devices, retraining employees, and investment in many other areas.

- Complexity – Convergence is probably the most complex initiative undertaken by service providers. The technologies enabling convergence is still not mature, the business model involves multiple external players resulting in complex matrix of unclear roles and responsibility, and finally, it is not clear which services will really work. In addition, new standards on convergence that are required for defining interoperability and interconnection between different types of protocols, devices, and network equipment are not mature yet.

- Transition – Many carriers around the world have already started migrating to an all-IP network, the most famous being BT's 21 CN effort. The challenge is not just in migrating from the existing network to the new all-IP network, but also in maintaining both of them in the interim. Many companies have capped their investment in yesterday's technology, but are not yet disposing of the old network. The existing environment is still supporting millions of customers and earning an enormous amount of money for the providers. For example, the voice market in the US alone is worth more than $100 billion, with most of this value still coming from the old network.

- Consumer confidence – Another key challenge to convergence is gaining public acceptance and confidence in this new way of life. Many people resist change and are skeptical of new technology. Service providers must ensure that the new forms of communication and entertainment are simple, manageable, secure and reliable. The quality of new services must be equal or superior to that of the old ones.

- Organizational – Convergence is completely overhauling the communication, media and entertainment industry. Silos within the industry and within a service provider's organization are going to be broken. There will be lot of consolidation, tie-ups within the industry, and re-organization within companies, which will lead to a lot of heartburn among employees. As with any major initiative, executive commitment is necessary. All changes that will be happening must be clearly spelt out, individual and company roadmaps must be charted, and new roles and responsibilities must be clearly defined. Employees must be trained in new ways of doing business and all necessary infrastructures must be in place. In addition to this, new processes and best practices must be defined so that the transformation becomes smoother.

- Regulatory – Current regulations on services is based on the presumption that a service is closely tied to a particular type of network. However, with network convergence, multiple services can be delivered over one network and that makes many of the existing regulations biased towards some services and networks. For example, in the US, voice service over IP was treated very differently from the voice service over the TDM network, thereby creating an imbalance in competition. Even today, in almost all countries the rules for TV transmission over cable are different from TV transmission over an IP network. This imbalance, although not easy to resolve, will hold back the growth of convergence. In addition to this, regulatory activities around standards, security, protecting consumer interests, intellectual property rights, and other public issues are not keeping pace with the developments in the marketplace.

16.7. Triple/Quadruple Play

Research about customer needs from all over the world has shown that customers like to purchase all their communication needs as packages or bundles from a single provider. Research has also shown that customer stickiness increases when they buy bundles from one provider as opposed to individual services from different providers. Taking into account the results of this research, service providers have started offering voice, broadband access, and TV services as a bundle, and such a bundle is commonly referred to as triple play. Many service providers like ATT have also added cell phone service to the mix, making it a quadruple play.

This concept of bundling together the top three or four basic services and selling them at a cheaper price is gaining momentum among all companies in the telecommunications industry worldwide. Bundling services is necessary for providers to stay competitive and remain in business.

16.8. Fixed Mobile Convergence

One of the hottest topics in convergence in recent years has been fixed mobile convergence (FMC) or fixed mobile substitution. The fixed phone line, synonymous with the telephone for several decades has suddenly come under threat from the mobile phone. Consumers around the world have been dumping their wireline phone for a mobile phone at the rate of 3–5% per year. To overcome this, wireline operators have devised a strategy whereby a single phone can act as a wireline phone when at home, and as a wireless phone while mobile. However, most of the recent efforts to accomplish a working FMC service have not produced expected results.

16.9. Convergence @ 2010

Although true convergence is not yet prevalent, but the seeds have been sown. The lines between fixed vs. wireless, wireline vs. wireless, and wireline vs. cable communications are blurring, with boundaries being tested and crossed. It is expected that by 2010, convergence will have completely broken the current silos within voice, video and data industries. These three industries will start to merge under one umbrella and companies will be fiercely competing with each other for a market worth a trillion dollars worldwide. Here is a brief list of the changes anticipated by 2010:

- The digital home, enabled by the PC, the set-top box, consumer electronics, network storage, and smart phone, will enable the seamless exchange of information like photos, emails, and all forms of content between devices and with the outside world
- The demand for services that personalize communication and entertainment, along with presence and location-based services, will be worth tens of billions of dollars
- Content, both professional and user-generated, will be king, but the broadband infrastructure enabling all this will generate a lot of revenue
- Place-shifting, enabled by convergence technologies, will allow the consumer to have their services and content follow them anywhere
- On the enterprise side, it is very likely that there will be a lasting IP wholesale transport business for all the existing telecommunication companies. There will be a surge in usage because of new productivity improving applications and services that bridge the distance gap required due to globalization.

By 2010, convergence will force the telecommunications sector to become much more diverse, dynamic, competitive, and consumer electronics-driven than it is today. The winners will be companies that embrace the change and move from being network-focused to customer-centric. The winners will be those that stop thinking of themselves as network controllers and instead become part of a value chain in which a variety of players will contribute to truly innovative applications.

17. Next-Generation Networks

As described in the previous chapter on convergence, one of the key technologies driving convergence is Internet protocol (IP). However, most of today's networks are based on a myriad of technologies, each capable of supporting only one type of service (Fig. 17.1). This is cumbersome to maintain and operate, difficult to introduce innovative applications, and finally, not economical at all.

Next-generation networking (NGN) is a term used to describe a new type of network that is completely based on IP technology. The general idea behind NGN (Fig. 17.2) is that one network supports all services (voice, video and data), which makes it cheaper and easier to maintain.

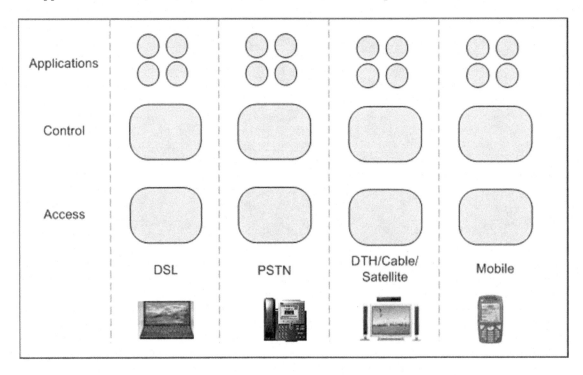

Figure 17.1 - Today's Networks

Many carriers are migrating to an all-IP NGN as it is the best way to deal with new realities like convergence, competition, deregulation, and customer expectation of more innovative applications and services at a lower cost. Many operators around the world are investing billions of dollars in the rollout of new IP-based networks. The most notable are BT's "21st-Century Network (21CN)", Korea Telecom's "Broadband Convergence Network" (BcN), and Deutsche Telekom's "Telekom Global Network" (TGN). Service providers around the world strongly believe that NGN will prove beneficial, as a single network will be capable of supporting all services (voice, video and data), and hundreds of next-generation applications. Consumers will also benefit through better and cheaper services, innovative new applications, and greater control and personalization.

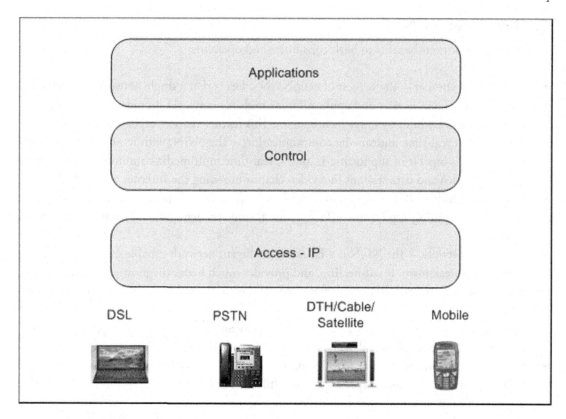

Figure 17.2 – IP Based Next Generation Network (NGN)

Migrating to an all-IP NGN is a gradual process. It is anticipated that in OECD (Organization for Economic Co-operation and Development – US, France, UK, Australia ...) countries, full wireline NGNs will be in place by 2012 and mobile by 2020, enabling the full convergence of voice, video and data services.

17.1. Characteristics of NGN

The next generation IP based network will be very flexible and more capable as compared to the current networks in place. The key differentiating characteristics are:

- Internet protocol (packet-based)
- Easily capable of scaling to support high bandwidth requirements
- Supports a wide range of services like VoIP and IPTV, and applications like MMS, photo share, caller ID on TV
- Capable of interacting with existing networks
- Capable of supporting convergence between fixed and mobile
- Capable of supporting multiple traffic types, with prioritization and different QoS levels to ensure appropriate performance for each application
- Has open APIs for third parties to build, own and operate applications
- Better security
- Uses minimum number of technology layers (predominantly IP and multi-protocol label switching – MPLS) resulting in simplified operations

17.2. Benefits of NGN

NGN brings enormous benefits to both consumers and operators:

- Simpler and cheaper – The biggest benefit NGN offers is that a single network is capable of providing the same services and applications as multiple networks do today. In addition, the network layers are built by fewer technologies. This further reduces capex and complexity
- Seamless and real-time multimedia communication – The NGN, with its singular IP-based technology, is capable of supporting seamless, real-time multimedia communication comprising voice, video and data sessions like video chat or browsing the Internet while talking
- High bandwidth – NGN is very much capable of supporting simultaneously multiple high bandwidth requiring applications IPTV, gaming, video on demand, and multimedia communications
- Intelligent network – The NGN is a far more intelligent network capable of re-routing during congestion/breakdown, is self-healing, and provides much better diagnostics in times of crisis. These characteristics will improve service, reduce downtime and makes it easier to operate the network
- Advanced applications – NGN is capable of supporting far more advanced applications like presence, location, profile, and Web 2.0 based value added services
- Improved customer experience – NGN allows advanced innovative applications that dramatically improve the customer experience and fulfill the thirst for innovative applications. These applications not only make life a lot easier, but also provide context-sensitive routing (e.g., if you are watching TV, all SMSs sent to your mobile are automatically routed to your TV) and context-sensitive help.
- Improved security – NGN networks have sophisticated technologies that are capable of providing a very high level of security.

17.3. Key Drivers

Almost every operator in the world is migrating to an all IP NGN. The key drivers behind such mass migrations are as follows:

- Operator needs – Operators have too many networks to support and maintain to provide basic services. It is expensive and there is no room to deliver the next generation of applications and services expected from consumers. With NGN, the operator will need only one network to provide all services and it will be possible to provide new innovative converged applications
- Consumer needs – Consumers are seeking innovative applications and services, more choices, attractive pricing, simplicity in billing, full mobility, and increased quality of service. The current crop of networks is incapable of fulfilling these requirements
- Enterprise needs – Enterprises are looking for simpler, cheaper, and versatile communication services. They want integrated voice, video and data services, flexible virtual private network (VPN) solutions, good video conferencing facilities, online collaboration tools, and better security to support an increasingly global enterprise. These requirements can only be supported by an all IP NGN
- Technology – It's not just in the last few years that operators have been seeking one network to support all services; there have been numerous attempts in the past, but almost every time the technology was not mature enough to support it. After being in use for more than 30 years,

IP technology has attained a very high degree of maturity, capable of overcoming the stress of supporting the requirements of multiple services on the same transport. In addition, technologies in the area of MPEG, IP multimedia systems (IMS) and service delivery platforms (SDP), FTTx, and ADSL, which were not available before, are contributing significantly to the growth of NGN.

17.4. Challenges

The shift towards next-generation networks (NGNs) began at the start of the twenty-first century. It is indeed one of the major transformations in the history of telecommunications. However, the transition is slow and fraught with many challenges, some of which are:

- Investment challenges – Transitioning to NGN requires an enormous amount of capital expenditure at all levels. There is no doubt that NGN brings considerable benefits in the long run, but shifting investment from current networks that are supporting key services to a new network that will take years to bear fruit is a difficult proposition
- Operational challenges – NGN is enabling service providers to offer a variety of new services like IPTV, mobile TV, fast broadband, and value-added services like location- and presence-based applications, multimedia messaging, unified messaging, dual phones, push-to-talk, and interactive gaming. However, providing fulfillment, assurance and billing for so many services will be very challenging
- Quality of service – Voice and video transmission requires a much higher degree of speed and reliability compared to data transmission. However, NGN is a packet-based network where all types of signals are treated equally. The signals are chopped into packets and routed as fast as possible towards the destination. This leads to delays during voice conversation or loss of clarity in video transmission, which sometimes is unacceptable
- Security challenges – Many components of NGN like gateways, routers, and call agents are computer technology-based products that are vulnerable to attacks from hackers. Addressing security threats and ensuring safety and reliability will be a daunting task
- Interoperability challenges – There are a variety of legacy networks that have been in operation for decades. But since NGN is new, it is expected to deal with interoperability with older networks so that communications across different networks is possible
- Regulatory challenges – Most of the regulations on services is based on the type of network used to provide it. However, in the case of NGN, one network is used for providing all services.

18.   Value-Added Services (VAS)

The telecommunications business has been changing rapidly in recent years, and nowhere are the changes more profound than in the services and applications that are being provided. For most parts of the 20th century, companies provided one or two services (for definition, see chapter 1 -section 1.3) and even fewer applications (for definition, see chapter 1 - section 1.3) within their lines of business and the world seemed content with it. However, new developments like convergence, next-generation network (NGN), standards, and IP multimedia subsystems (IMS) have now made it possible to create and deliver hundreds of new applications that involve not just communications, but also media and entertainment as well. In addition, unlike before where service providers created and delivered a handful of applications, today applications are being developed by a variety of players that includes large and small companies, and even individuals sitting in remote corners of the world. New applications are being developed and launched at an exponential rate to run over mobile, Internet, and TV. Collectively, these applications, whether they are run on voice, video or data platforms, are referred to as Value-Added Services (VAS). This term is a bit confusing as they are not really services but are applications, but it is a term used throughout the industry.

This chapter focuses upon understanding these new applications, the philosophy behind them, the reasons for their explosive growth, and what the future looks like for these new services.

18.1. Web 2.0

Web 2.0 is one of the primary concepts driving the innovation and adaptation of value-added services. The phrase was initially coined by Dale Dougherty in 2004 to signify the shift in the way the Web was being used. Even though Web 2.0 is a marketing term and dismissed by many pundits as just more dotcom hype, there is some substance to it.

Web 1.0, the first phase of the web, was mainly a tool for communication between one and many, where few posted information or provided services for the larger masses to consume. web 1.0 was mainly about enterprises setting up web sites to broadcast about themselves, perform commerce, and provide information like news and weather. The mode of communication was mainly from few to many. However, Web 2.0 is all about participation and contribution from all users, not just a few. Examples of Web 1.0 are commerce services like Amazon.com, information web sites like BBC.com, and company web sites like Microsoft.com. Examples of Web 2.0 applications are blogs, social networking Web sites like Facebook, Orkut, music-sharing services like Napster, content-sharing services like BitTorrent, and user generated content sites like YouTube. Each of the Web 2.0 applications is developed using the existing Internet infrastructure as a platform, and most are developed and launched by individuals sitting with a computer at home somewhere in the world.

Before we proceed any further, it's important to answer an obvious question. The impact of Web 2.0 on the web and publishing media is clear, but how does it affect traditional telecommunication companies? The name "Web 2.0" suggests that these services are mainly centered upon the Web, but this is just the tip of the iceberg. Web 2.0 application providers, regardless of which medium they are using today, are emerging as serious rivals to telecommunication companies. Today, more people are using social networking sites like Facebook and Orkut to communicate than old-fashioned way of calling people up or writing an email.

Companies like Joost (Internet TV) and Skype (voice) are not just offering competitive TV and voice services, but are increasingly looking towards using the Internet to offer tons of additional Web 2.0 applications on top of their base services. Facebook and Orkut are the new modes of communication. Google Map provides information about places of interest, negating the need for the yellow pages or the profitable 411 directory services provided by telecommunication companies. Web 2.0 based companies are creating an environment in which any user can participate, contribute, communicate and share. If telecommunication companies resist or fall behind in incorporating Web 2.0 concepts into their services, then they risk being left behind in a fast-evolving digital services value chain. The future for telecommunication companies is not about offering a few services, but of facilitating, aggregating and selling large numbers of Web 2.0 applications. It doesn't matter where and how the applications are developed or who owns and operates them; what matters is who holds this value chain together and makes the service available to consumers. For more information about Web 2.0, visit http://web20workgroup.com/.

Web 2.0 applications can be broadly classified into three categories:

- Social networking
- User-generated content
- Mashups

18.1.1. Social Networking

Social networking applications are supporting one of the most basic needs of human beings – to be social and connect with people. Earlier, voice communication over wireline or wireless service was the only means, but now there are many social networking applications capable of providing an alternative way of connecting with others. MySpace, Facebook and Orkut are not only some of the top social networking web sites, but also some of the most visited sites on the Internet. Social networking services allow users to find long-lost family and friends, put up their profile, share content, discuss, recommend, rate, provide feedback, tag, and broadcast events.

The concept of social networking has had an enormous impact upon the way we communicate and interact with others. More and more people, especially young people, are spending a lot of time online interacting and communicating with others. It has become an important alternative mode of communication, and a medium for people to have a voice in society. During the aftermath of the shooting at the University of Virginia, Facebook became a bulletin board for students to communicate and express their feelings. Existing modes of communication (mobile, TV, and wireline phone calls) are still relevant, but it is important to note that alternative ways and means of communication are emerging. Telecommunication companies around the world need to figure out how to incorporate these new means of communication into their Web, mobile or TV platforms and into services they are offering.

18.1.2. UGC

User-generated content (UGC) is unique content created by an amateur user as opposed to professional media producers, licensed broadcasters, and production companies. The content can be news items, video, or audio. Blogging, citizen journalism, video sharing, and pod casting are some of the most popular ways of generating and sharing content. This content is shared through Web sites like YouTube, Your News, Ohmynews and Technorati.

The advent of user-generated content marks a shift among some media organizations from creating content themselves to creating the facilities and framework for users to publish their content. User-generated content is one of the cornerstones of Web 2.0 and is extremely popular.

18.1.3. Mashups

The term mashup comes from the hip-hop music practice of mixing two or more songs. Mashups is an application available through the Web that combines content from more than one source in order to create an entirely new and innovative application. Typically, mashup content is sourced from a third party via a public interface like APIs or RSS.

There are many good examples of mashups that use Google Maps as one of their sources. For example, people have come up with applications that can show all the Indian restaurants in London, or spas anywhere in the United States, on Google Maps. For more examples, visit http://www.webmashup.com/.

18.2. Mobile 2.0

First, Mobile 2.0 is not about accessing Web 2.0 services through a mobile device. It starts with the definition of Web 2.0 – services driven by user participation, but adds the all-important feature of mobility. Mobile 1.0 was mainly about providing voice communication and some additional features like voice mail, but Mobile 2.0 is about a vast array of services that help people while they are mobile. A good example is getting the latest scores of your favorite sport on your mobile, uploading photos onto your mobile phone, or increasingly sophisticated applications that are capable of reminding you to pick up your laundry if you happen to be in the vicinity (location-based services) of your dry cleaner.

Mobile 2.0 is about services that go beyond those that are available through the web. Such services are being developed at an amazing rate, effectively knitting together Web 2.0 with the mobile platform to create something new. These services leverage mobility and are as easy to use and ubiquitous as the web is today.

Mobile 2.0 services will increasingly play a critical role in shaping societies around the world. According to the World Bank, mobile phones outnumber PCs by three times, and two thirds of the world's population lives within the range of a mobile network. Mobile phones are increasingly becoming fast and powerful, sporting color screens and cameras. In addition, the rapid penetration of wireless broadband access (WBA) technologies such as WiMax, 3G/UMTS is fuelling the growth of mobile phones around the world. There is a clear consensus among think tanks that mobile phones will become the most commonly used means of communication, and Mobile 2.0 services point the way forward for the mobile industry.

18.3. Impact on Operators

There have been various debates about the ultimate demise of the phone company. Earlier, all services were provided by the company that owned the network, but today due to the Internet, IP technology and regulatory changes, any company can create and sell services over a network. Network owners have gone from being the powerful sole service providers to just dumb pipe (network) owners. Does this mean the end of phone company as we know them? Many believe so, but this is not true.

Network operators still occupy an immensely important position in the value chain. They run the networks, including authentication, connecting calls, messaging, interconnect, roaming, and all the other complexities inherent in delivering seamless 24/7 uptime service. They still provide customer care services, manage billing, sell devices and coordinate among various service providers. Each of these activities is capital intensive and requires vast experience and financial might to build, support and execute them effectively. These activities cannot be performed by a new company run from a garage. In addition, each of the network operators still provides the basic voice, video and data services over which all the fancy new applications run.

Yes, the market dynamics have changed, but so have the phone companies themselves. They have transformed themselves from being the sole providers of all services (walled garden) to providers of a communication platform over which any outside company can provide a service (open garden phenomenon). If anything, traditional phone companies like ATT, Telstra, BSNL and BT have become more powerful and are showing a healthy rate of growth.

19. Mergers and Acquisitions

During the last decade, the telecommunications industry has experienced significant growth because of worldwide deregulation, liberalization, technological changes, and global market forces. On the other hand, the market also experienced a serious downturn during the beginning of the twenty-first century due to the dotcom crash. These upheavals forced the industry to merge, acquire, or build alliances (M&As) to survive. Hardly any other business sector has seen the same scale of M&A activity as the telecommunications industry. Barely a week goes by without news of another purchase, fusion or alliance.

This process of realignment has been going on for some time and there is no clear sign that it might end anytime soon. Recently in the US, SBC bought ATT ($16 billion), Sprint bought Nextel ($35 billion), Cingular Wireless bought ATT Wireless ($41 billion), Verizon bought MCI ($6.7 billion), and the new ATT bought BellSouth ($65 billion). In Europe, Sonera of Finland and Telia of Sweden merged to form the Nordic region's largest telecommunications company. This list could go on and on, with examples from almost every country.

The M&A wave has also affected other sectors of telecommunications. Lucent and Alcatel, the two big network equipment vendors, merged; European mobile phone manufacturers Nokia Corp. and Siemens AG have agreed to combine their telephone equipment units in a deal valued at roughly 25 billion euros ($31.6 billion). Even companies not in the telecommunications sector, like eBay (bought Skype) and Google, are entering the market. This is remarkable, as just a few years ago it was impossible to imagine eBay being in the telecommunications business. This shows just how volatile the once calm telecommunications business has become.

19.1. Why Merge?

With this madness continuing, it is important to step back and understand why this spate of M&As is happening. The following are some of the reasons:

- Consumer demand – The demand from customers (both households and enterprises) is clear: they need more for less and all services from a single source. This has led to carriers aiming to become fully integrated, one-stop, end-to-end global telecommunications and entertainment service providers. The aim is to provide all of the basic services (voice, video, data), but not all companies are capable of providing all the services by themselves. Thus, companies have started merging, acquiring, or striking alliances so that the merged entity can provide all the services. A classic example of a merger prompted by consumer demand for all services from one company is the merger between ATT and Cingular Wireless. Today, the combination of ATT and Cingular is capable of providing all basic services (wireline voice, DSL, VoIP, IPTV, wireless voice, wireless broadband).

- Deregulation – Deregulation in the marketplace has also broken down the silos of the traditional telecommunications and entertainment industries. The monopoly that traditional phone companies had in voice and cable companies had in the TV business are being broken down. Now, there are multiple choices for consumers to receive their communication and entertainment services from a number of players in the market.

- Competition – Competition has exploded and it is clear that only the most adaptable will survive. The telecommunications market has players from previously different sectors now essentially offering the same service. This has resulted in a large number of players turning the competition on its head.

Many of the M&As are designed to cushion the effects of these changes or exploit them.

19.2. Are M&As beneficial?

There have been serious concerns that mergers between the big telecommunication companies will lead to reduced competition and higher prices. This is a common notion, but if you dig a little deeper into the market dynamics, you will find that this is not entirely true. Rapidly changing technology is erasing the distinction between types of communication services and the companies that provide them. Today, a consumer can buy all their telecommunication and entertainment services from cable, or wireless, or satellite or wireline companies. Virtual network operators have also added to this competition. In spite of all the M&As, today's telecommunications market is still highly competitive. There are more wireless subscribers around the world today than traditional phone lines, data traffic exceeds voice traffic by a margin of 11 to 1. IPTV is fast becoming a viable alternative to cable and satellite TV; and VoIP is fast gaining popularity as a wireline voice service around the world.

Merged companies also have the intellectual and financial resources to spur innovation and propel the communications industry forward. New products and services that set the standard for how enterprises and individuals communicate will be put in place. The new merged entity will be capable of delivering the advanced network technologies necessary to offer integrated, high quality, competitively priced communication services to meet the evolving needs of customers worldwide.

19.3. Mergers @ 2010

There is no sign of a let-up in M&A activity in the telecommunications industry. Convergence in the industry, at the service and device level, along with the latest technology, is fuelling the fire of M&A. The one thing that will change by 2010 is the nature of the companies merging. Today, most of the mergers are between wireless and wireline or local and long-distance carriers, but the mergers of the future will be between telecommunication companies and media and entertainment companies. Such mergers have taken place (AOL and Time Warner) before, but that was a little premature. Today's market has the necessary demand in place for services from such mergers.

The other area where a lot of merger activity is expected is between transcontinental companies. Many countries in the developing world where the state held monopoly and a large stake in the telecommunications business have started loosening their grip. These new markets are extremely lucrative and unsaturated, which is enough reason for well-established enterprises from the developed world to move there.

Glossary

A

AM: Amplitude modulation – process of impressing information on a radio frequency signal by varying its amplitude

ANSI: American National Standards Institute – The U.S. standards organization that establishes procedures for the development and coordination of voluntary American National Standards.

ASCII: American Standard Code for Information Interchange

ARPANET: Advanced Research Projects Agency Network. A packet–switching network used by the Department of Defense, later evolved into the Internet

ARPU: Average Revenue Per User – a financial benchmark used to measure average revenue per subscriber

ADSL: asymmetric digital subscriber line – A flavor of DSL technology where the upload and download speeds are different

ATM: asynchronous transfer mode – A commonly used transport protocol

B

Bits – a binary digit which can take value of 1 or 0

Bytes – a group of 8 bits used to represents one character

C

CATV: Community Antenna Television or Cable TV

CDR: Call Detail Record – mechanism to record usage of service by a customer so that they can be billed

CDMA: Code Division Multiple Access – wireless standard for digital cellular phone service

CPE: Customer Premises Equipment – equipment installed in the customer's premises to receive services

CLEC: Competitive Local Exchange Carrier – a service provider in the US who provides competitive communications services in the ILEC's territory

D

dB: decibels – The unit for measuring the relative strength of a signal

DSL: Digital Subscriber Line – broadband technology over normal telephone line (copper wire)

DWDM: Dense Wavelength Division Multiplexing – an optical technology used to increase bandwidth over existing fiber optic backbones

E

ESS: Electronic switching system – hardware based electronic switching system

F

FCC: Federal Communication Commission – United States government agency that regulates telecommunication industry

FM: Frequency modulation – process of impressing information on a radio frequency signal by varying its frequency

Frame Relay – A commonly used transport protocol

G

GHz: Giga Hertz – A unit of frequency that is equal to one billion cycles per second

GPS: Global positioning system – a worldwide radio–navigation system developed by the US

GSM: Global System for Mobile Communications – wireless standard for digital cellular phone service

H

Hz: Hertz – A unit of frequency measured in cycles per second.

I

ILEC: Incumbent Local Exchange Carrier – primary service provider in the US operating within a local area

IN: Intelligent Network – telecommunications network architecture providing advanced functionality

IP: Internet Protocol – A packet–based protocol for delivering data across networks

ITU: International Telecommunications Union – A United Nations body involved in standardizing technologies and protocols

ISP: Internet Service Provide – a business that provides an individual with access to the Internet

K

KHz: Kilohertz – A unit of frequency that is equal to one thousand cycles per second

L

LAN: Local Area Network – A network connecting nodes of an entity within a small geographic area like university or office

M

MAN: Metropolitan Area Network – A network connecting nodes of an entity within a city

MHz: Mega Hertz – A unit of frequency that is equal to one hundred thousand cycles per second

N

NTSC: National Television Standards Committee – television signal standard used in the US, Canada, and Japan

O

OSI: Open Systems Interconnection – a suite of protocols, designed by International Standards Organization (ISO) committees, to be the international standard computer network architecture. Most of today's networks are based on OSI

P

PAL: Phase Alternating Line – television signal standard used mainly in Europe

PBX: Private Branch Exchange – a privately owned (usually by business) telephone–switching network

PSTN: Public Service Telephone Network – a network built by the world's wireline carriers

PTT: Push To Talk – a two way radio system using a mobile phone

Q

QoS: Quality of Service – the ability of a network (including applications, hosts, and infrastructure devices) to deliver traffic with minimum delay and maximum availability

R

RF: Radio Frequency – any frequency within the electromagnetic spectrum normally associated with radio wave propagation

S

SECAM: système electronique couleur avec memoire – television signal standard used in France, eastern European countries, the former USSR, and some African countries
STB: Set Top Box – used to receive and decode TV signal and deliver it to TV for viewing
SS7: Signaling System 7 – commonly used signaling system for voice communication

T

T1 Line – a high-bandwidth telephone line with a capacity of 1.544 Mbps, capable of carrying 24 voice channels
TCP/IP: Transport Control Protocol/Internet Protocol – suite of application and transport protocols that run over IP and used extensively in the Internet
TDM: Time Division Multiplexing – technology that transmits multiple signals simultaneously over a single transmission path

U

UHF: Ultra High Frequency – Frequencies spectrum from 300 MHz to 3000 MHz

V

VHF: Very High Frequency – Frequency spectrum from 30 megahertz to 300 megahertz

W

WAN: Wide Area Network – Network spanning cities

Can't find what you were looking for? Please refer the following websites:

http://www.atis.org/glossary/
http://www.its.bldrdoc.gov/fs-1037/
http://www.cellular.co.za/glossary.htm

Further Reading

Telecommunications is not a topic that can be fully understood by reading a book. The scope of the industry is vast and the dynamics of the market keep changing quite regularly. Following is a list of Website where you can find additional information like news, white papers, blogs, pod casts, analysis, etc:

General Topics

Telephony Online – www.telephonyonline.com
Telecom Magazine – http://www.telecommagazine.com/
Total Tele – http://www.totaltele.com/
TMCNet – http://www.tmcnet.com/
Network World – http://www.networkworld.com/
Light Reading – http://www.lightreading.com/

Broadband

Telephony Online – http://telephonyonline.com/broadband/
FCC – http://www.fcc.gov/cgb/broadband.html

IMS

Telephony Online – http://telephonyonline.com/ims/
TMCNet http://www.telecommagazine.com/2007/Insites/IMS/

Industry Forums

Alliance for Telecommunications Industry Solutions – www.atsi.org
Cellular Telecommunications & Internet Association (CTIA) – www.ctia.org
CDMA – http://www.cdg.org/
European Telecommunications Standards Institute (ETSI) – www.etsi.org
GSM – http://www.gsmworld.com/index.shtml
TMForum – www.tmforum.org
Wimax Forum – http://www.wimaxforum.org/home/

IPTV

Telephony Online – http://telephonyonline.com/iptv/
TMCNet – http://iptv.tmcnet.com/
Telecom Magazine – http://www.telecommagazine.com/2007/Insites/IPTV
Daily IPTV – www.dailyiptv.com

Regulatory bodies

CRTC (Canada) – http://www.crtc.gc.ca/
Federal Communications Commission (USA) – http://www.fcc.gov/
International Telecommunications Union – http://www.itu.int/net/home/index.aspx
Ofcom (UK) – http://www.ofcom.org.uk/
TRAI – (India) – http://www.trai.gov.in/

VoIP

Telephony Online – http://telephonyonline.com/voip/
TMCNet http://voipservices.tmcnet.com/

Wireless

Telephony Online – http://telephonyonline.com/wireless/
Telecom Magazine – http://www.telecommagazine.com/2007/Insites/3G/
Network World – http://www.networkworld.com/

Wimax

Intel – http://www.intel.com/technology/wimax/index.htm
Wimax.com – http://www.wimax.com/

Write to the author

Mistakes creep into any book despite the best efforts of the author, the copy editor and the production team. If you do find any errors, or have suggestions to improve this book, then please write to etm_reviews@cme-essentials.com. The author and the book production team would be highly obliged for you help.